SPECIAL STUDIES

The USAF Scientific Advisory Board

Its First Twenty Years, 1944-1964

Thomas A. Sturm

New Imprint by the
OFFICE OF AIR FORCE HISTORY
UNITED STATES AIR FORCE
WASHINGTON, D.C., 1986

FOREWORD

One of the highest compliments a USAF research and engineering officer can receive on his work is that it met the standards of "the Arnold-von Karman tradition." This allusion epitomizes objectivity of inquiry and thoroughness and excellence of performance.

This documented narrative traces that proud tradition from its genesis through the twentieth anniversary of its creator and most zealous guardian—the Scientific Advisory Board to the Chief of Staff and Secretary of the Air Force. Hopefully, the work will serve as both testimonial and concise source book on the invaluable contribution which this dedicated and uniquely-skilled companion-in-arms has made to the cause of American aerospace supremacy.

J. P. McCONNELL
Chief of Staff

HAROLD BROWN
Secretary of the Air Force

CONTENTS

APPENDICES—Continued

INTRODUCTION

General Henry H. Arnold told Dr. Theodore von Karman he needed him in the Pentagon during World War II "just to show the military that a college professor was good for something." Characteristically, there was more truth than hyperbole in Arnold's easy humor. Before the war, a near-disastrous gulf had opened between American men of arms and science at the top levels. This was Arnold's way of announcing that he intended to correct that mistake within the future Army Air Forces. The means he chose proved eminently successful. Dr. von Karman and the other distinguished civilian scientists who answered Arnold's call opened the way to an enduring partnership.

The nature of its assignment and the absence of precedence enabled the wartime Arnold-von Karman scientific advisory group to initiate procedures and standards which ultimately became unique hallmarks of its postwar successor. The wartime group formed, von Karman noted, "with the idea that we needed a future plan, a projection into the future." He accomplished the job by subdividing air science into its major parts and inviting the nation's foremost authorities in these technologies to join him. They met frequently in joint session to discuss general aims and progress but pursued their individual work independently. And their completed product—the prestigious multi-volume work entitled *Toward New Horizons*—went directly to General Arnold with conclusions and recommendations unaltered to fit any advisory group or Air Force preconceptions. When this group transitioned to the USAF Scientific Advisory Board (SAB) at war's end, its parts naturally converted to separate panels accustomed to having their findings forwarded directly to the Chief of Staff. At the same time, members favored continuing the wartime practice of meeting jointly to consider problems which cut across technological lines. Thus emerged the semi-annual general SAB meetings, which soon

developed into major forums of exchange for the daily prac-
titioners and part-time advisors who, together, comprised the
overall USAF research and engineering establishment.

The wartime group operated under explicit instructions,
but the postwar group underwent a long period of experimentation before it arrived at a satisfactory *modus operandi*. Perhaps the most salient lesson the SAB learned during these early
years was that it could not survive in the form it had already
come to regard as traditional unless it struck out on its own
when others lagged in soliciting its services. In this regard it
found it was no different from other Air Staff offices, despite
its unique nature. That is, had it contented itself with resting
on its laurels when assignments were not forthcoming, it soon
would have atrophied, perhaps expired. Hence there emerged
three prime initiators of SAB assignments: the Secretary and
Chief of Staff of the Air Force; Air Staff and field agencies;
and the board itself.

No matter how carefully official statements enunciated
board duties, they could not possibly encompass the SAB's full
role and significance. For example, Lt. Gen. Donald L. Putt,
while serving as board military director, offered the additional
observation that the board insured "that we in the USAF maintain the progressive outlook, that constant willingness to discard
the old and try the new." The SAB's technical director, Mr.
Chester N. Hasert, depicted it as "a unique organization for
quickly assembling the best scientific brains of the country
with a background in Air Force problems [and a] proven capability of obtaining quick answers to major policy decisions of a
technical nature which the Air Force would be slow to achieve
through other means." Dr. James H. Doolittle saw the SAB as
the organ through which "American science has an opportunity
to know Air Force problems and assist in their solution . . . a
public service of the highest order of importance." General
Nathan F. Twining, Air Force Chief of Staff, noted what he
regarded as the "job which in the long run is even more important than any of the others . . . the guidance [the SAB] can
give the Air Force in the field of fundamentally new ideas."
Dr. Clayton S. White, chairman of the Aero-medical/Biosciences
Panel, theorized that the Air Force did "most of the hard spade
work on the important research and development problems"
while the SAB served "as a sounding board and a mechanism
for refining thinking, [for providing] inputs which make for

analytical thoroughness, and [for bringing] balance and wisdom to conclusions and decisions through an impartiality not always evidenced by those who cannot see the forest for the trees." And Dr. H. Guyford Stever, in his capacity as board chairman, observed that the SAB served to spread "the idea of a positive approach to technology through the government and scientific community as a whole [and] that the indirect communications relationships of the SAB may well be one of its most important contributions to the Air Force and the Government."

Ample support for such estimates of the SAB's roles and accomplishments in the years 1944-1964 may be found in its studies and in official project histories. This volume seeks to mold data from these works and from the board's administrative records into a source book on SAB membership and organizational and operational turning points for these years. It should also make clear why, for example, the Hoover Commission in the mid-1950's adjudged the SAB to be "the best top structure for tieing in of science" it had seen within the government, and why Lt. Gen. James Ferguson, board military director, announced in 1964 that the SAB had "grown immeasurably" in importance to the Air Force in recent years and portended to become of even greater importance in the future.

The volume was written by Mr. Thomas A. Sturm of the Washington liaison office of the USAF Historical Division, Air University, Maxwell Air Force Base, Alabama, at the suggestion of Colonel Robert J. Burger, Secretary of the Scientific Advisory Board. The Secretariat and author are most grateful to the many former and present board members and associates who took time from their busy schedules to review the manuscript for accuracy and clarity prior to its publication. Requests for further information on matters of general Air Force content discussed in the volume may be addressed to the USAF Historical Division. Requests for additional information on specific board projects cited in the narrative or appendices or on the role of particular members in those projects should be submitted to the SAB Secretariat.

PART I

A SEAT FOR MINERVA ON THE AIR STAFF

Now I believe personally . . . that a Scientific Advisory Board should be effective now . . . . We shouldn't forget the more remote purposes and the desired projection into the future. . . . But certainly the voice of Minerva should be heard on the current problems.

———*Theodore von Karman,* speaking to the Scientific Advisory Board meeting of February 4, 1947.

CHAPTER ONE

NEVER AGAIN TO BE CAUGHT

I don't think we dare muddle
through the next twenty years the
way we have . . . the last twenty
years. I have worked with von
Karman the last twenty years, and
I was sometimes scared by the
knowledge he had that we weren't
using . . . . I don't want ever again
to have the United States caught
the way we were this time.

——Henry H. Arnold*

General Henry H. Arnold, wartime chief of America's Army
Air Forces (AAF), had full confidence in his hard-driving and
dedicated staff of World War II officer-scientists and engineers
at the Wright Field Engineering Division. Through a miracle
of innovation and adaptation they had produced the aircraft
and other weapons that he required for victory over the Germans
and Japanese. Now, in the summer of 1944, with victory in
sight, Arnold knew that this staff could also provide him with
valid plans for a suitable postwar AAF research and development program. At the same time, he felt he needed a plan
which looked far beyond the immediate period—a plan that first
examined thoroughly the latest scientific advances in the air
arms of all participants in World War II and then set forth
the future steps the United States should take to develop and
maintain the best air force in the world. To his mind, fitting
immediate plans into such a long-range blueprint was the first
essential step in guaranteeing his country's continued supremacy
in airpower.[1]

To help him get the "best brains available" to head the long-
range study project, General Arnold, as he related in his memoir
Global Mission, turned to his close friend Dr. Robert Millikan

---

*In address to Scientific Advisory Group, January 9, 1945.

See notes on page 179.

2

of the California Institute of Technology. What he hoped to find, he told Millikan, was a man with sufficient stature to attract to the project "practical scientists and engineers" who were expert in sonics, electronics, radar, aerodynamics, and other aspects of science that portended to influence future aircraft development.[2]

Arnold wrote that he and Millikan discussed the matter at some length. However, other evidence suggests that their final choice—Dr. Theodore von Karman, director of Caltech's famed Guggenheim Aeronautical Laboratory—was a foregone conclusion, subject only to Millikan's willingness to lose von Karman for a while and the latter's willingness to serve. Dr. Frank L. Wattendorf, former student and long-time associate of von Karman, afforded the best evidence for this view, noting that the scientist and the general had been close friends since the early 1930's when Arnold was a major in command of March Field, California, and von Karman the head of Caltech's Rocket Research Project (forerunner of the Jet Propulsion Laboratory). After he moved to Washington in 1936 as Assistant Chief of the Air Corps, Arnold retained a personal interest in von Karman's jet propulsion and rocket motor experiments, visiting the Caltech facilities many times. When in 1938 Arnold, now chief of his service, required technical counsel for overcoming opposition to his assuming control of research and development vital to the Air Corps mission, he solicited von Karman's aid. Their success, Wattendorf recalled, "broke loose the major facility construction and expansion of Wright Field, starting with the 20-foot, 40,000 horsepower wind tunnel, and encompassing all the laboratories." In 1940 von Karman accepted appointment as part-time consultant to Arnold and special advisor at Wright Field. Among his many contributions to air power about this time were his studies on the Bell XS-1 which later became the first manned aircraft to break the sound barrier. So confident did Arnold become in von Karman's judgment and counsel and so easily did they work together that from this time, according to Wattendorf, whenever Arnold needed a quick answer to a particularly tough scientific problem he often circumvented normal staff channels with a direct query to von Karman.[3] Thus, it appears logical to assume that he must also have planned to call on von Karman for the long-range study from the moment he conceived the idea.

In any event, Dr. Millikan sanctioned the move and Dr.

See notes on page 179.

von Karman accepted.* The initial appointment, issued October 23, 1944, read that von Karman would act as expert consultant on scientific matters relating to aeronautical engineering in the AAF, detailed to the Eglin Field Proving Ground Command in Florida. This enabled von Karman to enlist an initial cadre—consisting of Dr. Hugh L. Dryden, Dr. Wattendorf, and Dr. Vladimir K. Zworykin—to explore projects underway in the research and testing center at Eglin. Here, in a month-long stay, they established contacts for securing a flow of information on current research projects. They also witnessed launchings of "Chinese copies" of German V-1's. From their observations and discussions at Eglin they formed a general concept of the substance of the long-range study and drew up a list of the types of experts the group would need.[4]

Meanwhile, General Arnold formally established the group in a memo to von Karman on November 7, 1944, noting his conviction that the future security of the nation rested in part "on developments instituted by our educational and professional scientists."[5] Soon after, Arnold's deputy, Lt. Gen. Barney M. Giles, announced von Karman's appointment as Director of an AAF Long Range Development and Research Program.[6] On December 1, 1944, Giles announced the official establishment of the AAF Scientific Advisory Group (SAG), noting that it was attached directly to Arnold's office with the mission of assembling and evaluating facts on long-range research and development in the AAF and preparing special studies on scientific and technical matters pertinent to airpower.[7]

During this time, Dr. von Karman continued to build the SAG staff. In discussions with the members, General Arnold explained in greater detail what he had in mind. He asked them to forget the past, to use current equipment merely as a

---

*At the time of the Arnold-Millikan meeting, von Karman was in a sanatorium at Lake George, New York, which he entered in July 1944 to convalesce from an illness. Wishing to broach the matter directly but having little time to spare, Arnold asked von Karman to meet him at LaGuardia Airport where Arnold had a short layover during an official trip to Canada. Von Karman complied and they talked in an automobile on the airfield. According to Mr. Lee Edson, who assisted von Karman with his autobiography, von Karman was hesitant at first. He was not certain he should leave his work at the university or that he would fit into the Pentagon style of things. Arnold assured him of Millikan's favorable reaction, dispelling the first uncertainty. On the second, Arnold apparently assured him that if ever the occasion demanded he would personally see to it that the Pentagon fitted its style to von Karman's.

See notes on page 179.

point of departure for their boldest predictions, and to concentrate on manned and unmanned supersonic airplanes, smaller but more powerful bombs, air defense needs, communications possibilities, and all other phases of aviation that could affect "the development and employment of air power to come." In short, he wanted them to look 20 years into the future and prepare a workable guide for the air leaders who followed him.[8]

Organizationally, the SAG divided into two groups: permanent consultants who worked full-time in the Pentagon and others who continued at their regular employment but were on call as required. Dr. Dryden served as deputy technical director and Colonel Frederic E. Glantzberg as deputy military director. A secretariat of military officers handled administrative duties. Total manning eventually included some 30 civilian scientists, about a dozen military personnel, and a clerical staff.* At the outset, the full group—permanent and part-time consultants— met monthly to exchange views on their studies of where the AAF currently stood in relation to research and development possibilities and to formulate ways and means of proceeding with its long-range recommendations.

The full group met first on January 9, 1945, again on February 7, on March 7, and on April 3.† Dr. von Karman explained that their objective in these meetings included a search for ways "to secure scientific insight in a standing Air Force . . . to secure the interest of the scientists of the nation to help the future Air Force . . . and to educate the people of the nation that for our security we must have a strong Air Force."[9]

In late April 1945, Dr. von Karman and six SAG colleagues departed for Europe to familiarize themselves with the latest scientific thinking and to integrate this thinking into the AAF's future plans.[10] They spent two months abroad, interrogating top foreign scientists, including captured German scientists, and, in general, carrying out Arnold's wishes that they "observe, correlate and draw deductions from all possible enemy developments in being or under consideration . . . [and] of exercising imagination and scientific genius in recognizing possibilities which may develop from facts so collected."[11] On his return, Dr. von Karman gave SAG members an extemporaneous account of some of their experiences and findings. The trip was timed

---

*See Appendix J for the names of members and contributors to the work of the SAG.

†All of the SAG meetings convened in the Pentagon.

See notes on page 179.

well, he said, since VE-Day came a few days after they arrived in Europe making it possible for them to visit most places of value.   They had no electronics experts with them so their studies concentrated mostly on aerodynamics, missiles, and engines. They found Braunschweig, with its numerous laboratories for research in airplane design, ballistics, engines, and jet propulsion especially interesting.   The many German engineers and professors of aeronautical engineering who were still there had provided occupation authorities much information, especially on jet propulsion and guided missiles.   Von Karman noted that while much of the documentation had been destroyed at Braunschweig "95 per cent [of the data had been duplicated and] came out in very funny places—salt mines, wells, old shafts, or just buried in the backyard—and after some pressure and decent handling of the people, more and more of them came out every day." They also visited Gottingen, another primary aeronautics center.   While Braunschweig and Gottingen eventually went over to the British, United States authorities had them about six weeks and were able to provide the SAG with many important items, including microfilm records.   After Gottingen, von Karman and his colleagues visited England where the Royal Air Force briefed them on its progress with jet propulsion and missiles. Finally, they visited aviation laboratories and factories in Switzerland and Bavaria.*[12]

Many significant developments in American aviation had their genesis in SAG proposals submitted during and following the European trip.   For example, Dr. Dryden's missile report set the stage for much that was initiated in this area in the years immediately following the war.   On recommendation of Mr. George S. Schairer (forwarded by urgent cable from Volkenrode, Germany), B-47 design shifted from straight to swept wing.   Dr. Wattendorf's June 1945 recommendation for a new aerodynamic and propulsion center became the founding document for the Arnold Engineering Development Center project.   Finally, Dr. von Karman's timely recommendations sent

---

*Dr. von Karman also spent about two weeks in Russia on invitation of Russian scientists.   At one affair, he said, "all the professors of the military academy were there in general's uniform and all the big doctors in general's uniform."   He noted that the members of the academy got the highest food ration in the whole country.   "My feeling isn't that bread and meat should be the reward," he said, "but on the other hand I think it is quite a good idea, because in most of the capitalistic countries the people believe that a professor should lead a frugal life."

See notes on page 179.

directly to General Arnold beginning in May 1945 on the handling and processing of captured German scientific documents and equipment assisted greatly in reclaiming these invaluable tools for AAF postwar study and use.* Von Karman also successfully urged many prominent scientists to come to the United States and assisted in establishing programs (such as PAPER-CLIP) for actually getting them to their new homes.[13]

Dr. von Karman incorporated the findings from the European trip into a report titled *Where We Stand* which he submitted to General Arnold on August 22, 1945. He tried to show in this report "the main fields in which significant advances have been made and . . . 'where we stand' with some indications as to 'where we shall go.' " He also identified the following as new aspects of aerial warfare which he felt the AAF had to recognize as "fundamental realities" in future planning:[14]

> Aircraft, manned or pilotless, will move with speeds far beyond the velocity of sound.
>
> Due to improvements in aerodynamics, propulsion, and electronics control, unmanned devices will transport means of destruction to targets at distances up to several thousands of miles.
>
> Small amounts of explosive materials will cause destruction over areas of several square miles.
>
> Defense against present-day aircraft will be perfected by target-seeking missiles.
>
> Only aircraft or missiles moving at extreme speeds will be able to penetrate enemy territory protected by such defenses.
>
> A perfect communication system between fighter command and each individual aircraft will be established.
>
> Location and observation of targets, take-off, navigation and landing of aircraft, and communication will be independent of visibility and weather.
>
> Fully equipped airborne task forces will be enabled to strike at far distant points and will be supplied by air.

After examining how the United States currently stood in relation to German advances in such areas as supersonic flight, pilotless aircraft, and jet propulsion, he concluded that the "German achievements [were] not the result of any superiority in their technical and scientific personnel . . . but rather due to the very substantial support enjoyed by their research in-

---

*A similar SAG scientific search group, which Dr. von Karman was unable to accompany, went to India, China, and Japan in August-December 1945 on General Arnold's request. Their reports, which appeared as Far East Air Force Technical Intelligence Reports, also exerted considerable influence within AAF scientific circles in the immediate postwar years.

See notes on page 179.

stitutions in obtaining expensive research equipment, such as large supersonic wind tunnels, many years before such equipment was planned in this country."[15]

Upon completion of the *Where We Stand* volume, Dr. von Karman acted to expedite completion of their major project— the long-range study. The Japanese surrender in mid-August, and the emergence of a school of thought in the War Department that postwar long-range military research should be centralized and placed under civilian control similar to the way it was during the war made it necessary that they bring this work to a close as swiftly as possible. Opposed to the centralized concept, von Karman called on his colleagues at a late August 1945 meeting to complete their individual studies in short order so that the overall work could play a part in resolving "the question of how research and development should be secured for the services and how research and development should be divided between the Army and Navy on one side and the civilian agencies on the other."*[16]   To accomplish this, he suggested they greatly modify their plans. Originally, they had planned a main body "based on functional aspects of science in the Air Forces" followed by "a scientific analysis of the problems written as one text." In lieu of the textbook approach, he proposed a series of monographs introduced by a short summary volume.[17]

The SAG members adopted his proposal and, over the next three months devoted themselves to preparing the individual monographs while von Karman concentrated on synthesizing their broad recommendations into an introductory volume. In the first weeks of December, they rushed their labors to a close

---

*President Roosevelt asked Dr. Vannevar Bush, head of the War Department's Office of Research and Development during the war, to propose a postwar national scientific research program. After the President's death, Dr. Bush (in a report titled *Science: The Endless Frontier*, dated July 5, 1945) recommended creation of the National Science Foundation (NSF). As depicted in the report, the NSF would handle the major long-range military basic research projects, with the services performing just that research necessary to refine existing weapons. Dr. von Karman (as he noted in an interview with Air Force historian Mr. Samuel Milner in July 1960) immediately protested this view. Accordingly, General Arnold had Brig. Gen. Lauris Norstad discuss the matter with Bush. As von Karman recalled, Bush informed Norstad that he had been misunderstood, that the services could still carry out basic research on future weapons within logical limitations. Accordingly, von Karman and his colleagues proceeded on this assumption in their studies.

See notes on page 179.

and, on the 15th, von Karman presented to General Arnold the completed work—a series of 33 volumes titled *Toward New Horizons*. The first volume, written by von Karman, contained a discussion of the relation between science and aerial warfare, an analysis of the main research problems of the Air Force (from the point of view of its functions) and recommendations on organization of research. The 32 accompanying monographs treated detailed research programs in specific fields.[18] In them, von Karman later informed Arnold, the SAG scientists "attempted to combine a bold, forward-looking attitude with scientific soundness and appreciation of practical limitations."*[19]

In his introductory volume—titled *Science, the Key to Air Supremacy*—Dr. von Karman called attention to the increasing scientific and technological nature of warfare. Victory or defeat in the first world war had been decided mainly by human endurance. While the superiority of Allied tanks and the blockade of German shipping contributed greatly to German defeat, the main factor in the decision was "the complete exhaustion of human endurance on the German side." The second world war was far different, having from the beginning a technological character. Germany's overwhelming technological preparation, von Karman wrote,

> secured her first brilliant successes on the European continent. The shortcomings of the Luftwaffe in strategic bombing and the lack of experience of the German Army and its consequent poor preparation for amphibious operations caused the attack against England to be stillborn. The mounting tide of Allied, especially American, air power became finally the main factor in Germany's defeat. Even in the East, although the bravery and endurance of the Russians were perhaps the most important factors in stopping the German Army, the Russian march of victory to the West could not have been achieved without technological superiority, due partly to Russian and partly to American production.

Another new element in the second war, he noted, "was the decisive contribution of organized science to effective weapons." While science had played a role in all wars since time immemorial, never before had such large numbers of scientific workers "been united for planned evaluation and utilization of scientific ideas for military purposes." Thus, World War II had made obvious that future warfare would have a primarily scientific character, and *Toward New Horizons* attempted "to formulate some of the consequences of this conception for the [future] Air Forces."[20]

---

*See Appendix J for a listing of the studies.
See notes on page 179.

Dr. von Karman organized his introductory volume into three parts. The first—after dismissing the argument currently held in some quarters that atomic weapons would negate the need for large military forces in the future—concluded that among its future tasks the Air Force would have to reach and hit remote targets swiftly and with great power, secure air superiority in any future war, transport large forces of men and arms swiftly to any point on the globe, and defend the United States against other air forces. Von Karman made the point here that "only an air force which fully exploits all the knowledge and skill which science has available now and will have available in the future, will have a chance of accomplishing these tasks." The second part distilled and synthesized the key principles and proposals contained in the supporting monographs and sought to estimate future Air Force research and development needs in relation to missions. The final part summarized the SAG's recommendations on the "organizatory character" of future Air Force research and development requirements—the fundamental principles which should govern the organization of Air Force research, the extent to which the Air Force should cooperate with scientific institutions and industry, the facilities that the Air Force would require, and the scientific training of Air Force officers.[21]

Throughout this first volume, von Karman stressed time and again his conviction that the future Air Force had to be equipped—physically, intellectually, and psychologically—to gear operational planning to scientific development. He warned that it would not be possible to relegate scientific problems and officers to one niche and military problems and officers to another, noting that "scientific results cannot be used efficiently by soldiers who have no understanding of them, and scientists cannot produce results useful for warfare without an understanding of the operation."[22] He charged Air Force leadership with the task of creating and maintaining a climate of mutual respect and cooperation between the scientists and military planners, enjoining them to remember "that problems never have final or universal solutions, and only a constant inquisitive attitude toward science and a ceaseless and swift adaptation to new developments can maintain the security of this nation through world air supremacy."[23]

General Arnold distributed copies of the report among Air Staff members in early January 1946, hailing it as "the first

See notes on page 179.

of its kind ever produced" and an excellent guide for research and development planning in the coming years.[24]   In May 1946, Air Materiel Command (AMC) chief Lt. Gen. Nathan F. Twining, having been asked to evaluate and comment on the report, said he fully endorsed "the basic principles of the responsibilities of the Air Forces in the scientific domain" set forth in Dr. von Karman's introductory volume. These stipulated that the Air Force (1) had fundamental responsibility for insuring that the nation was prepared to wage effective air warfare, (2) had to call on all talents and facilities in the nation and support the development of facilities and creative work of scientists and industry, (3) required the means of recruiting and training personnel who had full understanding of the scientific facts necessary to procure and use equipment which was more advanced than that used by any other nation, and (4) had to be authorized to expand existing Air Force research facilities and create new ones to perform its own research and also to make such facilities available to scientists and industrial concerns working on Air Force problems.   These particular principles were fundamental, General Twining said, and if the Air Staff implemented just this much of the report initially the AAF would have laid a sound foundation for the future.[25]

By mid-1946, AMC and Headquarters AAF staffs were hard at work drawing up detailed plans for implementing the salient recommendations of the report.[26]   The fact that approval for many of these plans was either delayed for long periods or never forthcoming did not detract from either the value or the reputation of the report.   The timeliness of its appearance, the impeccable reputations of its authors, the objectivity and directness of its approach and language, and, above all, the genuine and inspiring claim it made on all Americans, civilian and military, to share the task of keeping America supreme in the air assured its immediate and lasting success.   It became the lodestone and the touchstone for Air Force research and development, a final arbiter of argument, a main source for inspiration and motivation. One top Air Force figure ascribed the report's enduring reputation to "the unqualified confidence and support" which it engendered from the start among scientists and industry in Air Force research and development.[27]   Another noted 20 years later that in re-reading the report at that time he was astonished at the validity of its predictions.[28] From such tributes are legends formed, and as long as there remained a U.S.

See notes on pages 179 and 180.

Air Force the documents listed under the title *Toward New Horizons* promised to stand among the other respected pillars of that service's heritage, tradition, and pride.

CHAPTER TWO

A FEW BRICKS FOR THE
FOUNDATION

> I believe the Air Force is a high
> building constantly adding new
> construction. If we can put in a
> few bricks, especially in the lower
> levels which carry the weight—
> if we can help build the founda-
> tion—we will be very happy.
>
> ——Theodore von Karman*

Having finished the work for which they had come together
and with the war done and demobilization the order of the day,
the Scientific Advisory Group prepared to disband. The members met for the last time in the Pentagon on February 6, 1946,
where General Arnold thanked them for their services and expressed the hope that they, and their colleagues in the universities and industry, would help the Air Force "continue its advance and preeminence" in the peacetime years ahead.[1] Maj.
Gen. Curtis E. LeMay, recently appointed to the new office of
Deputy Chief of Air Staff for Research and Development, also
addressed them, inviting their continued interest in the scientific problems of the postwar Air Force. On March 1, Dr. von
Karman resigned his government position, formally ending the
project.[2]

Meanwhile, a seed planted in the summary volume of *Toward New Horizons* for establishing a scientific advisory group
on the peacetime AAF staff had already taken firm root. Herein
Dr. von Karman and his colleagues had proposed the formation
of a permanent advisory council of eminent civilian scientists
to report directly to the AAF commander on important techno-

---

*During discussion at an April 7, 1949, SAB meeting.

See notes on page 180.

logical developments and promising scientific research. They recommended that men invited to serve on this council be thoroughly familiar with the work and needs of the Air Force but have their main interest outside the Air Force, be "experts with broad experience in the various branches of science involved," and provide a cross section of the nation's scientific thought.[3] Dr. von Karman later personally endorsed and expanded on the proposal to General Arnold, indicating his "strong belief" that the AAF commander would need frequent and valid scientific advice, and a group of distinguished scientists on ready call could best provide it. As von Karman first envisioned the group, it would consist of 10-15 part-time consultants supported by a full-time staff of a military director (preferably a brigadier general), a civilian scientist or an officer with scientific training to serve as secretary, and clerical personnel. One of the scientists would serve as chairman and spend several days a month with the full-time staff.[4]

General Arnold had circulated the proposal among his staff for comment.[5] Although fully supporting it, Arnold nevertheless felt that with his retirement imminent his replacement should decide whether or not to bring civilians of the high order contemplated into intimate involvement with the Air Force's daily affairs. If Arnold had any serious doubts on the matter, his generals quickly dispelled them. As General LeMay expressed it, the wartime SAG "rendered such signal service to the Army Air Forces during the war that it has made obvious the necessity for continuation of such a service as an essential part of Headquarters staff planning."[6] Accordingly, LeMay and von Karman met to work out the details for activating the group.*[7]

They submitted their plan to General Carl A. Spaatz (soon to replace General Arnold) on January 9, 1946. It differed from von Karman's earlier ideas on several key points. First, it proposed a considerably larger group calling now for a chairman and 30 members. Also, it did not provide for a military director as such but called on the Deputy Chief of Air Staff for Research and Development to serve as ex-officio member, perform liaison with the Air Staff, provide one of his civilian

---

*Much credit for the rapid transition of the group from a wartime to peacetime structure belonged to Colonel Roscoe C. Wilson (later Lt. Gen. and board military director) who represented General LeMay in the detailed planning for the changeover.

See notes on page 180.

scientists as the group's secretary, and furnish clerical and other administrative support. Finally, it aligned the membership into panels devoted to specific technological areas.[8]

On February 13, 1946, exactly one week after the wartime group's terminal meeting, General LeMay took the first step toward activating the permanent group by requesting permission to transfer the wartime group's functions, and civilian position allotments to his office.[9] The request was approved and the move completed on February 28.[10] Meanwhile, Dr. von Karman and his former SAG colleagues had drawn up a recommended membership, to whom General Spaatz on March 14 issued formal letters of invitation. In soliciting their assistance, Spaatz noted that "the success of the Air Forces in the recent war was due, in large measure, to the integration of our scientific, industrial and military resources [and] future security will, in turn, depend on the degree to which we are able to continue this intimate, constructive relationship." The new group—to be called the Scientific Advisory Board (SAB)— would help to insure the survival of this relationship by affording the Air Force "guidance in the planning and programming of research and development activities."

During the next three months, membership on the new SAB was established and, on June 17, 1946, it convened for the first time. On all major counts, it was manned, structured, and administered according to the von Karman-LeMay plan. Membership totalled 30, including von Karman who had accepted General Spaatz' invitation to be the first chairman. Of the 30, over two-thirds had served on the wartime group. The rest were equally regarded in their fields and equally familiar with military research and development needs.*

The members spent the first two days of their week-long first meeting in the Pentagon establishing their organization and procedures and becoming acquainted with the overall postwar AAF program. In his welcoming remarks General Spaatz assured them that the AAF intended to "pay close attention to all your advice which we hope will be as critical as you can make it in order to keep us on the right path."[12] Dr. von Karman then explained appointment and tenure policies and the panel structure. Orginally, he and General LeMay had wanted to

---

*See Appendix C for names and tenure of office of SAB members from 1946-1964. Appendix E indicates members' panel assignments for those years.

See notes on page 180.

limit tenure to two-years, with one half of the initial board receiving only one-year appointments to allow the start of an annual 50 per cent rotation. In the end, however, they had invited all members to join for one-year only, abandoning the rotation plan for fear it would be too restrictive. The policy adopted set no limitations on the duration of membership. Board officers would review the roster annually and re-issue invitations to those individuals whose services the AAF continued to require and who were agreeable to remaining aboard.[13]

All but Dr. von Karman were assigned to one of five panels. The Aeromedicine and Psychology Panel had four members with Dr. W. Randolph Lovelace II as chairman. The Aircraft and Propulsion Panel had seven experts in airplane and engine design, propulsion, and materials under Dr. C. Richard Soderberg's chairmanship. Eight experts on molecular and nuclear energy-producing materials under Dr. Lee A. Dubridge's chairmanship staffed the Fuels, Explosives, and Nuclear Energy Panel.* Five guidance authorities under the leadership of Dr. Dryden made up the Guidance of Missiles and Pilotless Aircraft Panel. The Radar, Communications, and Weather Panel contained experts from each of these technological areas with Dr. Zworykin as chairman. The arrangement was far from perfect, Dr. von Karman pointed out, possessing several incongruities and overlappings. But it was a start and necessary modifications would be made in the future.[14]

In answer to some members who feared that the SAB might degenerate into five seminar groups concerned only with items within the bounds of their disciplines, Dr. von Karman pointed out that assignment to a single panel was not intended to restrict members of different panels from working together.

---

*Dr. von Karman explained the rationale for joining these subjects into one panel at this time as follows: "From a scientific point of view . . . there is no great difference between electronic reaction, which is called molecular, or between nuclear reaction. And from the practical point of view, I think it is advantageous to put these topics together because the whole procedure in atomic development is yet in flux, so it is perhaps too early to make one separate panel for molecular energy only. I thought it would be better, for the time being, to get together nuclear energy with explosives and fuels. Also many questions are similar. After all, the questions of terminal ballistics, the science of destruction, even if the scale is different, are similar for ordinary bombing and bombing by atomic bomb. Also, other questions, for example, theory explosion waves, have some scientific foundation and require the same methods of investigation in both cases."

See notes on page 180.

General LeMay agreed, intimating that he felt most of the problems anticipated at that time would work themselves out as the SAB gained experience. "The panels as organized," he said, ought to be able "to handle the best of questions that we will be asked to solve. However, there may be some come up that fit into no particular panel, and possibly another committee will have to be appointed at the time or the board can consider them as a whole body and work on them."[15]

Board management was vested in an Executive Committee consisting of Dr. von Karman, General LeMay, and the panel chairmen. General LeMay and his staff would call problems to the attention of the Executive Committee who would then assign them to a panel, a special SAB committee, or even to outside scientists. The Executive Committee would meet every second month; the full board twice a year; and the panels as necessary.* Dr. Ralph P. Johnson, civilian scientist in General LeMay's office, was designated to serve as the interim SAB secretary pending assignment of a permanent secretary.[16]

Following the two days at the Pentagon, the members flew to Wright Field where, on June 19, General Twining and his AMC staff discussed both the fiscal year 1947 AAF research and development program and the broad aspects of future planning and then provided an inspection tour of the physical plant. In the following two days they met as panels to receive detailed briefings and as a board to draft reports on the five-day meeting.[17]

On August 19, Dr. von Karman forwarded the finished report to General LeMay. Consisting of a summary backed up by detailed panel papers, the report included among its many findings an endorsement of a draft directive prepared by LeMay's office for implementing within the AAF the War Department's recently enunciated policy on research and development in the postwar services. Von Karman hailed the new policy, which separated research and development contracts from procurement contracts, "as a very important step for utilizing scientific talent and facilities available throughout the

---

*The board never instituted this ambitious Executive Committee meeting plan. For the first years, the Committee convened immediately prior to or during the semi-annual full board meetings and in occasional emergency session. Later, as noted in a subsequent chapter, meetings were stepped up to four per year.

See notes on page 180.

country for solution of scientific and technical problems facing the Air Forces."* They strongly supported the AAF requirement for a special air engineering center (which in 1950 became fact as the Arnold Engineering Development Center) including the construction of facilities there for research in transonic and supersonic speed ranges.†

The SAB also concurred in the current and planned distribution of research and development funds, recommending only that meteorological research be given a larger share. They approved Wright Field's close collaboration with industry but suggested that AMC offices rely less on industry program suggestions and more on their own researches, particularly in the aerodynamic and propulsion areas. Finally, they noted that the reforms proposed in *Toward New Horizons* for the acquisition and administration of military and civilian technical personnel in the AAF had not been implemented. It was important, Dr. von Karman reiterated, that these men be given the opportunity to keep pace with the programs of scientific research through study in civilian institutions and opportunity to do individual research in AAF laboratories.[18]

General LeMay forwarded the SAB report to General Spaatz on August 29, recommending that Spaatz approve and authorize its implementation. Spaatz did so on September 4, adding, however, that the recommendations, had to be carried out "within current budgetary and Headquarters AAF policy limitations."[19] Thus General Spaatz had confirmed his pledge to back the board but it was equally clear that under existing circumstances the AAF would be hard-pressed to enact any recommendation which required extra funds or extensive revision of current forces.

Meanwhile, the SAB members had dispersed and, despite their ambitious planning during the June conference, failed to convene either by Executive Committee or panel for the remainder of 1946. The simple truth was that no one called on them to do anything. In late December 1946, Dr. von Karman

---

*This eventually became Air Force Regulation 80-4. One key Air Force leader later remarked that the document "would have probably remained in coordination stage for many more months were it not for strong support from the Board."

†General Twining later said that "the leadership of Dr. von Karman and your Board in the early days of the Arnold Engineering Development Center was perhaps the major factor in the establishment of that important facility."

See notes on page 180.

checked on the current state of affairs. He found that Dr. Johnson (and Mrs. Marie Roddenberry, who served with the original SAG then transferred to the SAB as its only full-time employee and performed services far transcending her assignment as administrative assistant) had compiled Air Staff and AMC comments on the August report. This had been the extent of activity, however. Von Karman complimented them on their work to General LeMay but noted that such a small staff could not possibly hope to adequately handle board affairs. He asked that a proper secretariat be established soon, to include as a minimum, a senior officer "for contact and coordination with AAF and outside agencies," a junior officer or civilian with a scientific background for handling technical matters, and a clerk to relieve the administrative assistant from stenographic and typing duties.[20]

Soon after, Dr. Johnson put the matter even more strongly to General LeMay, noting that the AAF had to decide soon how it was going to handle and employ the SAB or risk demeaning or even destroying its potential value. Calling the board together once or twice a year, presenting it with problems and inviting comments, then placing it on call for consultation between sessions caused the AAF little work and still gave the AAF, in Dr. Johnson's words, the "apparent benefit of advice from a group of experts as to the health of the research and development program." But this practice had some serious disadvantages, too. Unless board members kept continuously informed on AAF activities and plans—which they had not been able to do since their June meeting—they would have only an imperfect background for judging the adequacy of programs examined in their infrequent periodic meetings. Also, by failing to observe a regular procedure for bringing specific problems to SAB's attention—which had been the case since the June meeting—the Air Staff was likely to fail to solicit advice when it could be most available and helpful. Finally, Dr. Johnson warned, if these practices persisted, board members, finding their role relatively passive, might suspect that they were being used merely for ornament—that the AAF was primarily interested in citing SAB before the Congress or Bureau of Budget and less concerned with help the board could give. However false this opinion might be, it could obviously do the AAF much harm if it grew among board members and they passed it on to their associates.[21]

See notes on page 181.

Dr. Johnson pointed out that von Karman's ideas for employing the board called for more effort on the AAF's part, but were in keeping with the Air Staff policy set forth in early 1946. This statement had envisioned "continuous active cooperation between the AAF and the Board, with a flow of general information and specific problems from the AAF to the Board and a flow of suggestions and advice from the panels and individual members of the Board to the AAF."[2] Admittedly, money and manpower were currently very hard to get. But AAF statisticians had estimated that the Board, as currently operated, cost only about $60,000 per year and the AAF would be money ahead if it paid the additional nominal amount required to give the board adequate administrative support. If this could not be done, Dr. Johnson concluded, then perhaps the board should be considered a "dangerous luxury" and dropped rather than run the risk of exposing the AAF to charges that the SAB was mere window-dressing and only a convenient, prestigious means for rubber-stamping what the AAF wanted in the way of new facilities and equipment.

Though he did not say so directly, Dr. Johnson obviously favored the second plan of action. "The Board members are not essentially of the elder statesman type," he said.[22]

Their collective opinion on broad policy questions is valuable as a sample of the opinion of the country's scientific and technical talent on such questions, but their chief potential value to the AAF lies in their background of detailed knowledge which can be applied to detailed scientific and technical problems confronting the AAF. The best utilization of this asset would occur if each Board member were personally acquainted with the men in the AAF who are concerned with his particular field of competence, and could keep contact with these men and their problems by direct correspondence and occasional visits. The regular meetings of the board would then be an occasion for bringing the collective wisdom to bear on general policy questions; the SAB office in Headquarters AAF would monitor and assist the Board-AAF cooperation but would not need either to control it or to keep it stimulated artificially.

Acting on von Karman's and Johnson's urgings, General LeMay in February 1947 requested an additional officer allocation for his office to serve as SAB secretary, asking that the incumbent have a scientific or technical background and experience in AAF research and development.[23] He was successful in his bid and Major Donald M. Alexander assumed the position in April.

See notes on page 181.

Meanwhile, the board met for the second time on February 4-5, 1947. The meeting followed the same general pattern as the first, with the same merits and defects. Internally, SAB officers filled the few resignations submitted and expanded the membership slightly to staff a sixth panel called Weather and Upper-Air Research, under Dr. Henry G. Houghton's chairmanship. The panel with which Weather formerly had been linked became the Electronics and Communications Panel and Dr. Dubridge accepted its chairmanship following Dr. Zworykin's resignation from the board. Finally, Fuels switched to the Aircraft, Propulsion *and Fuels* Panel, which remained under Dr. Soderberg's chairmanship, and the former panel became simply Explosives and Nuclear Energy with Dr. Robert H. Kent replacing Dr. Dubridge as chairman.

As in their 1946 meeting, SAB members received Air Staff briefings on AAF research and development plans then, in plenary and panel sessions, framed their report on these plans which Dr. von Karman sent to General LeMay on May 20, 1947.[24] Two of their recommendations concerned queries which General Spaatz personally posed in late 1946. One dealt with the development of a continental air defense system. Since it would be a costly undertaking, Spaatz said, the AAF wanted an air defense system which met immediate needs yet was flexible enough to allow continuous upgrading in step with latest technical advances. On SAB's recommendation, the AAF asked RAND for a comprehensive study of this subject. On Spaatz' second request, which concerned the extent of AAF responsibility for research and development of systems affecting AAF operations, Dr. Detlev W. Bronk asked the National Research Council, of which he was chairman, to study a portion of the problem while Dr. William J. Sweeney solicited the viewpoint of private industry on the matter.

Other major SAB recommendations called for (1) increased emphasis, to include actual experimentation, on missile guidance systems, (2) continued emphasis on integration of missile and warhead designs "especially in the initial formulation of the military characteristics themselves," and (3) continued emphasis on advanced research projects. On the latter point, Dr. von Karman noted that in the face of current budget reductions "it may seem natural to eliminate the most visionary and most advanced ideas, on the basis that such research takes a long time to bring results and because they make it difficult to meet

See notes on page 181.

the obvious needs of the immediate future." He advised the AAF to keep in mind, however,

> that while the more conventional ideas have a chance of survival without the active support of the AAF—because of their applications to commercial aviation, and to other armed forces—the really visionary and advanced ideas must depend upon the AAF as the only source of support. This point of view should be kept in mind in the difficult situation of stretching inadequate funds over a wide field. It can be successfully applied only with the utmost efforts in coordination with other agencies.

Generals LeMay and Spaatz approved the report and directed that appropriate agencies act on it insofar as funds and policies permitted.[25]

The AAF had again demonstrated its pleasure with the board's proceedings, but the board itself did not share this view. As Dr. von Karman explained, "no work was foreseen for the members of the board [and] . . . most of the members felt that they were called in to learn something and then give their approval to a program already completely prepared before it was presented to the board, without the cooperation of the board."[26] Dr. Johnson indicated that his impression was about the same: members were generally dissatisfied with meeting sporadically for a smattering of information and a chance to render off the cuff opinions and wanted to be more useful.[27]

Originally, Dr. von Karman had planned a second meeting in 1947, but the involvement of top officers in problems arising from unification of the services followed by the need to organize the new U.S. Air Force caused him to cancel the meeting. By September, he felt sufficiently discouraged by the lack of board activity during these months to write Major Alexander that "if we shall continue . . . the members should be asked to make some positive contributions."[28] By early 1948, as noted later, he was even more discouraged. Fortunately, the new USAF had reached a point of stability where it could commence to accord the board the attention and support it required if it were to survive.

Much of the credit for this increased recognition belonged to Lt. Gen. Laurence C. Craigie. On October 10, 1947, in the Air Staff reorganization which followed unification, the position of Deputy Chief of Air Staff for Research and Development was abolished and its functions, including SAB administration, assumed by the new Directorate of Research and Development under Craigie.[29] In the change it appeared that the SAB had

See notes on page 181.

been demoted one echelon in the Air Staff hierarchy, since Craigie's office was one of several directorates within the Deputy Chief of Staff, Materiel. However, subsequent events proved this was neither the intent nor the effect of the change.

From the moment he assumed office, General Craigie evidenced his determination to seek greater employment of SAB's talents. Early in January 1948, he indicated to AMC his concern over the fact that "the services of the board in the past have not been utilized to the maximum advantage." Either persons at the research and development working level were not familiar with the SAB's functions or they mistakenly assumed that the board existed solely for the personal use of the Chief of Staff and that all matters referred to it had to originate with him. Craigie informed AMC that the SAB wanted to be put to better use and invited questions or problems that the board could assist on. Maj. Gen. F. O. Carroll, AMC's Director of Research and Development, agreed with these observations. He suggested that the board issue a statement of policy, indicate the type of work it could handle, and detail procedures for seeking its assistance. He also pointed out that the Air Staff statement affixed to past SAB recommendations that they had to be implemented within current budgetary and policy limitations had made it "impossible for the board and AMC to achieve the more worthwhile advances desired."[30]

Soon after, General Craigie informed Dr. von Karman that he was "aware of the dangers of having the board considered a 'showpiece' or 'rubber-stamp' for already established programs and policies" and informed him of his exchange with General Carroll. He then suggested that the board, as a start toward resolving these problems, issue an official statement of purpose and explain its manner of operation.[31]

On March 17-18, 1948, von Karman and Craigie called a third board meeting in the Pentagon with the primary intent of finding ways and means to energize the organization. Dr. von Karman did not mince words in informing the members of his discouragement over the past months. "When I arrived here and talked with General Craigie, I told him that because we had a group which perhaps contributed something—not very much—to the problems of the military establishment, it does not necessarily follow that it should be continued or should be further developed if they (the military establishment) do not feel it is necessary and if we do not feel it is necessary." However

See notes on page 181.

Craigie had convinced him that the Air Force now, more than ever, desired to retain the board's guidance and help. On reflection, von Karman now saw the goals of the SAB as somewhat different than they were a few years ago. "I believe that there is a necessity for coordination of what I would call the 'normal scientific life' in the research laboratories of the universities, and one which meets the needs of national defense—to find means which allow a combination of both activities," he said. In amplification of this point, he said he believed "that we are here to do what you might call 'crystal gazing' . . . and attempt to foresee what will happen in ten years. Some of the problems the board should embrace are . . . current scientific problems. We should also establish a certain procedure which will make it possible for the military establishment to use the services of the individual board members for urgent problems."[32]

Proceeding along Dr. von Karman's guidelines, the membership, in the March 1948 meeting, initiated a host of significant procedural changes. Agreeing that they had been too passive in the past, they formed several standing committees and directed them to study various key Air Force projects which they felt required expert advice. Dr. William R. Sears (chairman), Dr. Nicholas J. Hoff, Prof. Courtland D. Perkins, and Dr. Wattendorf formed one committee to study both the Arnold Engineering Development Center (AEDC) interim plan and the personnel policy for staffing the center. Drs. Houghton (chairman), Nathan M. Newmark, and Louis N. Ridenour formed another to review the organization and functions of the newly-established basic research office at Wright Field.* A third committee of Drs. Hoff (chairman), Sears, and H. Guyford Stever set out to work with AMC and Headquarters USAF to examine the proper organization and utilization of technical intelligence. Drs. John P. Markham (chairman), Kent, and Duncan P. MacDougall formed a fourth committee to plan the establishment of a Society for Military Sciences. A committee of Drs. Sears (chairman), Dr. Hoff, Prof. Perkins, and Drs. Hsue-shen Tsien, and Wattendorf agreed to draw up a recommended program for future use of the XS-1 research aircraft. Finally, a sixth committee of Drs. Dubridge (chairman), Pol E. Duwez, Joseph Kaplan, Irving Langmuir, and Prof. Perkins addressed themselves to an evaluation of the adequacy of Air Force administration of Project RAND.[33]

---

*See note, page 31.
See notes on page 181.

To make SAB services more available to field agencies, the Executive Committee prepared for circulation a listing of members willing to serve as special consultants and sanctioned their invitation to AMC and other Air Force agency meetings. The only proviso placed on the procedure was that the agencies would forward to the SAB secretariat a copy of the invitation and a report of the findings of the meetings. In this regard, SAB officers subsequently met with AMC laboratory and division chiefs and discussed the procedure in further detail.[34]

The members also acted to eliminate the criticism that board meetings in the past had given too little time to panel discussions and to problems at the working level. For the next meeting they decided to assign the first day to individual panel discussions with interested Air Force officers. The morning of the second day would be devoted to panel reports to the full membership. In the afternoon they would hear presentations from Air Force representatives or SAB members on matters not covered in the earlier sessions. When circumstances required, they would extend meetings into a third day but only those members with a direct interest in the subject matter to be discussed would attend.[35]

Finally, members recommended that the board's roles and missions be widely publicized primarily through publication of an Air Force directive and concurred in the selection of Dr. Dryden as deputy SAB chairman.*[36]

Two other major defects in SAB administration now remained, but these required General Spaatz' direct concurrence before the board could eliminate them. One concerned SAB's organizational place on the Air Staff. Although formed to serve the Chief of Staff, the board appeared on the chart as a function of an office at the directorate level. While this might seem a picayune point, General Craigie noted that it sometimes led "to considerable confusion within the Air Force, the military establishment, and among the board members themselves concerning the location and status of the board in the organization of Air Force Headquarters." The organization chart

---

*In explaining the establishment of this position to the members, General Craigie noted that it would put the board "in a little better position" on those occasions when SAB representation at important Air Staff meetings was desirable and the chairman could not attend. After agreeing that Dr. Dryden was "a very acceptable individual," they broached the subject of his appointment "late this noon at luncheon where he had little time to defend himself." The position was subsequently renamed "vice chairman."

See notes on page 181.

ought to reflect the relationship between the board and the Office of the Chief of Staff, he felt, both to maintain SAB prestige and confirm the Air Force's high regard for it. Moreover, "those agencies . . . concerned with research and development would recognize the fact that the recommendations submitted by the board are of significant importance and consequently merit thoughtful consideration."[37]

In early April, Dr. von Karman put the problem a little more directly to General Spaatz, observing that "a board of this nature can be of maximum value to the Air Force if it is a board of the Chief of Staff and is responsible directly to the Chief of Staff." Von Karman also broached the second administrative defect at this time—the continued lack of a military director. His solution was to appoint the Director of Research and Development to this position inasmuch as this officer was "in contact with all scientific institutions and organizations, both military and civilian, and is consequently immediately aware of the scientific problems with which the Air Force may have to cope and which should be brought to the attention of the board for their advice and recommendations." Recognizing that this was not in keeping with "proper channels of command," von Karman recommended that the Director of Research and Development, while acting as SAB military director, have direct access to the Chief of Staff.[38]

In response, General Spaatz invited his deputy, General Hoyt S. Vandenberg, General Craigie, and Dr. von Karman to discuss these and other board affairs with him on April 15, 1948. At this significant meeting, Spaatz concurred fully with their recommendations. The SAB would be part of the Office of the Chief of Staff, the Director of Research and Development would be the military director and report directly to the Chief of Staff on SAB matters, and all board recommendations would go directly to the Chief of Staff, a point never precisely established earlier.[39]

Later that month, Maj. Gen. William F. McKee, Assistant Vice Chief, announced the transfer of the SAB secretariat to the office of the Chief of Staff and impressed on all offices and agencies the importance of cooperating with and aiding the board in its mission.[40] On May 14, 1948, the terms of the Spaatz-von Karman understanding received official promulgation in Air Force Regulation 20-30.*

---

*See Appendix I for an account of the major provisions of AFR 20-30, as amended and revised over the years.

See notes on page 181.

CHAPTER THREE

CONFIRMING THE PARTNERSHIP

> In my experience, [the Ridenour
> Report] is the first civilian report
> that has ever been acted on by
> a military organization when they
> didn't have to. They are usually
> filed in the wastebasket.
>
> ——James H. Doolittle*

The secretariat received its own manning table for the first time concurrent with its transfer from the Directorate of Research and Development to the office of the Chief of Staff in April 1948.†¹ In the summer of that year, Lt. Col. Teddy F. Walkowicz replaced Major Alexander as SAB secretary. Shortly after, Lt. Gen. Donald L. Putt succeeded General Craigie as Director of Research and Development and SAB military director.††

In the fall of 1948, Dr. von Karman and the new SAB military officers reviewed the record of the board to date with the purpose of identifying and removing any further obstacles to efficient operation. They concurred in recent suggestions that general board meetings ought to concentrate less on formal briefings and more on creating an atmosphere in which members and Air Force leaders, engineers, and scientists might communicate informally. This would permit the SAB to become more intimately acquainted with spheres of activity where its unique capabilities could be applied and, hopefully, give the board the

---

*In remarks at a meeting of SAB officers with the Air Staff, January 30, 1950.

†Initial manning of the secretariat consisted of one military officer, a civilian administrative assistant, and a civilian typist.

††Colonel Walkowicz served as a military staff member on the wartime Scientific Advisory Group. He subsequently served as board member (1959, 1961-1962) following his resignation from the Air Force.

See notes on pages 181 and 182.

reputation it would like to establish "of being of the greatest service and of the smallest nuisance to those who benefit by [its] activities."[2] They used this approach in the November 16-18, 1948 meeting, inviting key Air Force research and development specialists and allowing them time for a frank and informal exchange of information and viewpoints. The procedure proved highly satisfactory and became the hallmark of most future board meetings. As Dr. von Karman observed, "by having eliminated the window dressing, so to speak, of a lot of speeches or presentations and having really gotten down to where the pick and shovel boys do their work . . . we have gotten more out of it than in prior meetings."[3]

A second matter of concern was that too many Air Force officers still were not convinced of the propriety of having "a bunch of civilians tell the Air Force what to do." The new SAB regulation (AFR 20-30) had helped to dispel some of this notion. The board officers now proposed to give the SAB secretary some assistance to free him for further educating the rank and file of the Air Force on SAB functions and objectives. Too, they felt that SAB members could help by striving to "brush wings" often with high-ranking Air Force members and so better understand "the plans and apprehensions—even the daydreams—of high staff officers about the future." To assist in establishing this rapport, they thought that SAB officers ought to sit in on key Air Staff meetings and also meet occasionally with top officers at informal luncheons and dinners.[4] These proposals, too, were soon acted on. In the spring of 1949, the secretariat received permission to employ a civilian assistant secretary and acquired Mr. B. J. Driscoll's services. And, from 1949 on, luncheon meetings between SAB members and the Chief of Staff and other top military officials became a regular feature of SAB procedure.

In their final and most significant finding, von Karman, Putt and Walkowicz agreed that "a large reservoir of potential utility of the SAB to the USAF remains untapped" and that to date SAB impact on the developing Air Force was slight. The major problem continued to be that SAB recommendations, after approval by the Chief of Staff, were dispensed for action without indication where this should take place. And, as noted earlier, the approval always carried the budgetary and policy

See notes on page 182.

restrictive provisos.* Obviously, if approved SAB recommendations were to succeed, they had to be "accompanied by a definite indication of the level of authority on which action should be taken," and AMC had to be freed to at least seek alteration of funding and policy limitations.[5] On this problem, General Vandenberg, the newly appointed Chief of Staff, gave his personal support in early 1949. After approving the SAB recommendations of the November 1948 meeting, Vandenberg notified his deputies and the AMC commander that they were to use these "in all cases . . . as guides in long-range USAF planning" and advise him personally whenever a strong SAB recommendation was not or could not be implemented. He also directed each deputy to henceforth furnish the SAB a summary of actions taken on all its recommendations.[6]

The new procedure did not ensure that budgetary and policy restrictions would no longer thwart the implementation of SAB recommendations. "Even the objective advice of an eminently qualified group which has only our best interests at heart cannot, for practical reasons, always be followed immediately," Vandenberg said.[7] However, the new procedure did fix responsibility on top officers either to act on SAB recommendations or to apprise the Chief of Staff and the board chairman of the problems which prevented such action.

Two additional important changes occurred in SAB procedures in late 1948. In one, Lt. Gen. Benjamin W. Chidlaw, successor to General Twining as AMC commander, established an office on his staff specifically as a point of contact on SAB affairs. Its duties were to distribute and follow up on SAB reports, seek SAB assistance on special AMC problems, and coordinate with the secretariat whenever AMC invited SAB members to serve as consultants.[8] In short, it further advanced the recent drive to establish greater rapport and understanding

---

*They cited two consequences of this procedural weakness. In one instance, SAB had recommended that the AMC Power Plant Laboratory do a small amount of basic research despite the serious budgetary restrictions currently hampering such work. In other words, the proposal clearly called for a readjustment of the Air Force budget to give more funds to the project. However, it went to AMC with the usual "within current budgetary and Headquarters policy limitations" restriction. This, of course, ended the matter. In another instance, the SAB recommended a reorganization of the AEDC and its staffing with competent scientific personnel. Headquarters USAF tacked on the same restriction and sent the recommendation to AMC where it eventually ended up in the hands of an engineer at the working level as a reference document.

See notes on page 182.

between SAB and AMC personnel. Dr. von Karman expressed his pleasure on this step, noting to board members that "heretofore you have been something nebulous so they couldn't quite visualize how you were going to help them. I think they felt a little strange about this high-powered group that was going to criticize them and tell them how to do their job. I think we have gone a long way toward eliminating that . . . [and] are going to see much improvement in the future."[9] General Chidlaw concurred, noting that AMC was "fully aware of the potential value of the SAB . . . and most anxious to realize this potential to the fullest extent."[10]

In a second change, the Air Force for the first time offered contracts to SAB members to reimburse them for board or AMC consultant duty. To date, members had received only per diem and travel expenses. Now, each would receive a contract for $50 (in addition to per diem and travel) for each day of consultative service, including that performed at home. General Chidlaw deemed the innovation a progressive move, feeling the program would "enhance the value of the Board to the USAF as [among its other advantages] it will permit sending . . . studies . . . to interested members for thorough analysis prior to board considerations."[11] Dr. von Karman accepted the change "with some hesitation, and only with the provision that individual board members could refuse compensation if they so desired."[12] Since it continued in effect over the ensuing years, the change obviously proved a satisfactory one, enabling the Air Force to feel free to solicit the services of many members, particularly those from the universities for whom board and special consultative duty might otherwise have worked a financial hardship. About half the board members accepted such payment during the first year (fiscal year 1949) and over the following years.

In summary, the many changes introduced in late 1948 and early 1949 effected a rejuvenation in SAB procedures and spirit. The consensus was that for the first time since it formed in 1946 the board was geared, administratively and conceptually, to become the organization visualized in its charter.[13] Needed now was a major assignment whereby the SAB might demonstrate the value which its officers and the top command were certain it offered. Such an assignment was not long in coming, emanating in large part from the report of the SAB's November 1948 meeting. Many of the recommendations in the report

See notes on page 182.

concerned the need to improve Air Force research and development facilities and practices, including several for enhancing the prestige and influence of the research and development staff in Headquarters USAF.*[14] As General Putt later described the nature of the problem, it was part and parcel of the same battle which "Dr. von Karman and the SAB have been fighting . . . since 1945 when he and General Arnold personally took the matter up."[15] After General Vandenberg read the November 1948 report he met with von Karman and Putt to discuss the problem.

As a result of the report and the conference, General Vandenberg decided to call on the SAB at its spring 1949 meeting for a comprehensive review of Air Force research and development. When, at the last minute, he was called to an urgent meeting of the Joint Chiefs of Staff, Vandenberg asked his deputy, General Muir S. Fairchild, to present his request. At this meeting on April 7, Fairchild read them the talk Vandenberg had intended to deliver personally. "The United States Air Force is well aware that continued technical superiority is one of the vital decisive elements in modern air power," Vandenberg wrote. "I am determined that our research and development activities shall have adequate support in funds, facilities, and properly-trained personnel, and that the USAF shall continually increase the efficiency and effectiveness of our

---

*Earlier in 1948 the SAB convinced the Air Force of the need for an office devoted exclusively to basic research. As noted by General Craigie (in a December 1965 note to the SAB Secretary) the Navy had had for many years the Office of Naval Research which "deservedly enjoyed a fine reputation." Dr. von Karman and General Craigie, beginning in the summer of 1947, strongly supported creation of a similar agency on the Air Staff. Though no one disputed the need for the office, AMC felt it ought to control the function. That command's wishes prevailed and, in February 1948, an office with basic research as its primary duty was established within AMC's Engineering Division. In February 1949, as the Office of Air Research, it was moved from under the Engineering Division to a position paralleling it with Lt. Gen. (then Colonel) Leighton I. Davis at its head. Though the new office failed to achieve the degree of independence which the SAB had visualized it was a significant step toward an improved program. And, as such, General Craigie noted, "it was an important contribution to Air Force research on the part of Dr. von Karman and the SAB." (An excellent account of the actions taken over the years on behalf of creating an effective USAF basic research program may be found in the Headquarters, Office of Scientific Research history for January-June 1964, Chapter II, *Organizing for Research, 1944-1951: The Founding of the Office of Scientific Research*, by Mr. Nick Komans).

See notes on page 182.

development work on new aircraft, missiles, and air defense systems."[16]

Of the SAB-USAF relationship, Vandenberg said, "We have hurdled many difficult and pressing operational problems [together] during the few years since the end of the war." He now proposed that they "take an equally critical look at our equally-important long-range technical objectives." Specifically, he asked the board to give him its "frank and objective advice" for drawing up an "ultimate plan" for Air Force research and development facilities. The plan, he said, ought to explore every facet of the problem. He particularly wished to make certain "that personnel and administrative policies and practices are adopted which will insure that our facilities are given proper leadership [and are] staffed by competent military and civilian technical personnel."[17]

Dr. von Karman expressed the board's gratitude "to hear the needs of research and development so clearly defined by the highest command in the Air Force" and assured General Fairchild that the SAB would "go into these difficult problems very carefully."[18] Later that month, von Karman reiterated to General Vandenberg his pleasure "over the opportunity to contribute toward the solution of problems which will influence substantially the development of the Air Force . . . and national security for many years to come." He noted that he and General Putt had "explored thoroughly" the various possible methods of handling the project and had decided to create a small working group of SAB members and other prominent experts.[19] As finally selected, the group consisted of two SAB members—Dr. Ridenour, who accepted the chairmanship, and Dr. Wattendorf—and seven non-SAB members—Dr. James G. Baker, Dr. James H. Doolittle, Dr. James B. Fisk, Dr. Carl F. J. Overhage, Dean Ralph A. Sawyer, Prof. John M. Wild, and Mr. Raymond J. Woodrow.*

Officially designated the "SAB Special Committee on Research and Development Facilities, Budget and Personnel," but quickly dubbed the Ridenour Committee, the group convened for their first meeting in the Pentagon on July 11, 1949.† General Vandenberg greeted them, expressing his pleasure at

---

*Drs. Baker, Doolittle, and Overhage later accepted SAB membership.

†Dr. Ridenour noted at one meeting that he preferred to regard the group as the Doolittle Committee, with himself serving as chairman of the Doolittle Committee.

See notes on page 182.

getting this "outside help in things that we are not competent in ourselves" and noting that "we realize, probably more than even you think, that we lack many aspects of the kind of help you can give us." He also informed them that he had given the Air Staff *carte blanche* "to get the answers to any of the problems that you people want within our competence." The Air Force, he said, had "no traditions or any inhibitions, because we are a new Department [and] we would like to start off research and development on the proper foot, and I think that with the advice and assistance of this group we should be able to do that."[20]

General Vandenberg then proceeded to "get a few things off [his] chest," as he expressed it, concerning the status of Air Force research and development and possible measures to strengthen it. As he saw it, after World War II, "everything went down hill so fast that the first thing we had to pay attention to was to get a sort of fire-bucket brigade ready in case something should break." He believed the Air Force had done well, meeting the Russians head-on during such crises as the Berlin blockade and managing to carry out the tasks at hand despite funding, personnel, and equipment handicaps. In short, he said, "we have gotten our people together now to the point where we feel that we have a force-in-being; therefore, our thought naturally turns to 'where do we go from here?'" This question had three important facets: (1) how could the Air Force acquire and retain capable scientists, (2) how should development be fitted into plans and operations, and (3) what was a proper distribution of research and development funds. To help decide these issues he wanted the Ridenour Committee to "give us a picture of what we ought to be doing but what we are not doing. I think that you can decide best how to present the problem so that we know what the problem is and how we can lick it, [and] . . . if it contains the answer that we are after, we will carry on from there."[21]

The committee devoted the next six weeks to exploring the problem, meeting about twenty times at a dozen Air Force and other military and government centers across the country.[22] Dr. Doolittle (in what Dr. Ridenour dubbed his "dandy little pep talk") emphasized why it was so important that the Air Force build up an adequate research and development capability as quickly as possible and generally described the tenor of the committee's approach. "I feel that the only thing that is going to keep us out of war is our technological advantage," Doolittle said. "It is far better to keep out of war than to win a war. If we permit a potential enemy to get ahead of us technologically . . . that is the surest way to start a war. I feel that the time has come to make some sacrifice from today's continuing emergencies in order to prepare for tomorrow's eventualities—to jar loose some funds, some competent personnel from the daily requirements in order to prepare for tomorrow's requirements.[23]

The committee finished its report in mid-September and Dr. von Karman sent it to General Vandenberg on the 21st. Organized into ten chapters introduced by a summary section and supported by six appendices, the study as one general officer later described it, "really summarized in beautiful English some of the feelings that many of us had had for some time . . . that we needed a new organizational emphasis laid upon [Air Force research and development]."[24] In Dr. Ridenour's words, the study's major recommendations called for "the establishment of a Research and Development Command separate from the Materiel Command, and . . . some reorganization . . . in the [Air Staff] to set off the function of research and development from the logistics—procurement, mobilization, and supply—functions." These recommendations were predicated on the committee's belief, Ridenour continued, that "the time has now come to put part of the effort available to the Air Force—the word 'effort' comprising the usual things, namely men, money and work—put part of the effort not on the Air Force in being, the Air Force of today, but on the Air Force of tomorrow."[25]

The committee refrained from getting too deeply into the details of reorganization, purposely choosing to leave this to a board of officers meeting under Air University auspices. The reasoning here, General Putt explained, "was that where this committee is very competent to determine what is wrong with research and development . . . and can point out possible solutions or ways that they think solutions might be achieved, [they are very reluctant] to tell the military that this is the way they will draw their organization chart."[26] The committee also offered only general answers to Vandenberg's questions on facility development and on the methods for "insuring the most effective interaction between technical development, on the one hand, and plans and operations on the other." They entailed too large an assignment for the committee to pursue in detail within its time allocation.[27]   Subsequently, in accordance with Dr. von Karman's promise at this time, other SAB committees and panels pursued these subjects.*

The report circulated throughout the Air Force and evoked an instant and not surprising furor.   General Putt observed that "it started some very deep thinking," demanding "a new concept, a new religion, on the part of those people who are in the top positions that have been making the final decisions which have vitally affected research and development in the Air Force."[28]   The culmination came on January 23, 1950, when the Air Force established the office of the Deputy Chief of Staff/Development on the Headquarters USAF staff and the Air Research and Development Command (ARDC).[29]   The Directorate of Research and Development and the Directorate of Requirements (from DCS/Operations) transferred as the major elements of the new major staff office.   Maj. Gen. Gordon P. Saville assumed the new deputy position, and General Putt remained the Director of Research and Development.

For the SAB, this complete acceptance by USAF top officials of its recommendations on this critically important issue, an acceptance which Dr. Doolittle hailed as unprecedented in military-civilian relations, dispelled any lingering doubts members may have had as to their value to the Air Force.†   The report also spread the SAB's name and purpose among a gratifyingly large Air Force audience.

In May 1950, General Vandenberg asked the SAB to do a complementary study on Air Force medical research and development.   He suggested that the group formed for this purpose seek "to determine whether we were doing a proper job, whether we were organized to carry out this job, and, particularly, . . . whether the existing organization was geared for the new and larger responsibilities we would have as an independent Air Force."[30]   Because several of the persons most competent to perform this study were out of the country, the SAB delayed action on the request.   The committee finally chosen consisted of Drs.

---

*Dr. von Karman appointed a Facilities Committee in late 1949 with Dr. Markham, chairman, and Drs. Hoff, Lovelace, Stever, and Wattendorf members, replacing the AEDC standing committee (see Chapter Two) which had played such an instrumental role in the establishment of AEDC. Dr. von Karman explained that problems concerning the AEDC could be handled henceforth by ARDC and that the SAB Facilities Committee would expand its scope of concern to include all USAF facilities.

†See Dr. Doolittle's comment at the beginning of this chapter.

See notes on page 182.

Lovelace, chairman, Edward J. Baldes, Donald W. Hastings, Kaplan, and Shields Warren, who were SAB members, and six non-SAB members.*

Convening for the first time on October 9, the Lovelace Committee subsequently investigated the state of Air Force medical programs and resources within the U.S. and submitted a report in December 1950.[31]   General Vandenberg directed that it be implemented the following March.[32]   As comprehensive in scope as the Ridenour Report, the Lovelace Report soon became the bible of Air Force medical research and development planners.   The buildup of personnel staffs and facilities for promoting research, teaching, and improved practices in aerospace medicine which took place in the early 1950's owed much of its direction to this report.

Meanwhile, the board acted to improve its own capability to treat problems falling within the overall category of "human resources."   In 1949, the Aeromedicine and Psychology Panel invited several social scientists to join it, changing its name to the Aeromedicine and Social Sciences Panel at this time.   However, as the Lovelace Report soon confirmed, the clinical aspect of aerospace medicine was of sufficient magnitude of itself to demand the full attention of one panel.   Consequently, after examining several alternatives the SAB decided to relegate strictly clinical matters to an Aeromedicine Panel, under Dr. Lovelace's chairmanship, and to create a separate Social Sciences Panel.   In February 1950, General Vandenberg invited Mr. Charles Dollard to chair the new panel, noting that it would assume cognizance over such areas as sociological, social psychological, and cultural anthropological programs, non-clinical psychological research, research in psychometrics, aptitude and proficiency tests, training, and training devices, military management, leadership, morale, psychological warfare, and strategic intelligence.[33]   This action increased the number of SAB panels from six to seven.   The number rose to eight soon after when the Aircraft, Propulsion and Fuels Panel divided into two under the chairmanship of Dr. Sears (Aircraft) and Dr. Soderberg (Fuels and Propulsion).

---

*Of the non-members, Drs. Loren D. Carlson, Paul M. Fitts, and John B. Hickam later accepted board membership. The others were Drs. R. Lee Clark, Jr., Magnus I. Gregorson, and John H. Lawrence.

See notes on page 182.

PART II

FROM DREAMS TO ENGINEERING PROBLEMS

When we started these activities in the beginning of 1944, General Arnold emphasized that he did not want the board to be concerned with any of the current projects which at that time were being carried out for immediate purposes. He wanted us to look into the future. Now since 1944 and 1945 several changes have occurred . . . the things that we were talking about . . . were mere dreams at that time, but they are facts now. Many of these vague ideas have since become engineering problems.

——*Theodore von Karman,* in speaking to members of the Scientific Advisory Board Executive Committee, March 1948.

CHAPTER FOUR

OUT OF CRISIS A TRADITION

> The board's effectiveness, at the
> highest policy-making level in the
> Air Force, is the result of our joint
> recognition of the ever-increasing
> importance of science to Air War
> . . . . I have long watched the re-
> sults of your studies take effect in
> the Air Force. . . . When you act as
> a group and recommend on major
> questions that require my personal
> action I will, of course, always
> be interested in your opinions and
> give them my most serious personal
> attention. Together in this partner-
> ship . . . we must achieve the vital
> task of providing the Air Power
> which stands between us and de-
> struction.
>
> ——Nathan F. Twining*

Whatever fancies the United States may still have har-
bored in 1949 that it could reduce its military forces to their
traditionally small cadres and continue to rest confident that
its freedoms were adequately safeguarded disappeared after the
Russians successfully tested a nuclear device in August of that
year. The test dispelled another chimera: the Russians were
not the scientific unsophisticates many had supposed them to
be.† Even those who had faced up to the harsh reality that the
nation's supremacy in these weapons would not go unchallenged
for long were surprised by the rapidity with which the Russians

---

*In address to the SAB, October 19, 1953.

†As Dr. Doolittle expressed it (in an interview for the USAF Video
Historical Documentation Program in June 1965), until this time the world
had mistakenly viewed the Russians as "agrarians with long beards who
went about with their shirttails hanging out their trousers."

unlocked the atom's secrets. Clearly, the United States now had to push its own weapon development timetables ahead or risk coming out second best in the deadly race for nuclear supremacy.

The SAB's immediate reaction to the Russian nuclear success, as Dr. von Karman noted at the November 1949 meeting of the SAB Executive Committee, was "to get people who were more intimately connected with atomic problems."[1] Soon after, General Fairchild involved the board directly in the crisis by calling for an emergency session and requesting assistance in strengthening the nation's air defenses. Until the Russian atomic test, the Air Force had frequently expressed concern but not undue alarm over its inability to create even a minimal air defense of the nation's people and industries. It was doubtful that anyone, including the Russians, would dare launch an air attack so long as the Strategic Air Command retained its overwhelming nuclear retaliatory advantage. Now that the advantage was disappearing, the Air Force felt obliged to press harder for air defense. First, however, it needed a sound plan of action, and General Fairchild asked the SAB to help draft such a plan.[2]

SAB officials discussed the issue at length and, on November 29, 1949, Dr. von Karman forwarded their recommendations to Generals Fairchild and Vandenberg who promptly approved them.[3] Accordingly, the SAB formed the Air Defense System Engineering Committee (ADSEC) and assigned it the task of developing "equipment and techniques—on an air defense system basis—so as to produce maximum effective air defense for a minimum dollar investment." The committee also set out to "help determine quantitative, factual data concerning current and future operational techniques and equipment" and, hopefully, suggest means that "would help improve the operational effectiveness of the existing Air Defense Command." Since ADSEC would work closely and frequently with an experimental unit of the Air Force Cambridge Research Laboratories, the SAB staffed it with "eminent scientists who could conveniently assemble regularly and on short notice, at that facility."[4] Subsequently, Dr. George E. Valley accepted chairmanship, with Dr. Allen F. Donovan, Dr. Charles S. Draper, Dr. Houghton, and Dr. Stever as members. Two non-SAB scientists, Dr. John Marchetti and Dr. George C. Comstock, joined them.

See notes on page 183.

Beginning their work in December 1949, the Valley Committee "worked diligently and with considerable success" for the next two years.[5]   At the peak of their labors, members met every Friday with government and Massachusetts Institute of Technology (MIT) scientists at the Cambridge facility. As Dr. Valley described their operations, they functioned as informally as possible, making most of their recommendations verbally to the Air Force officials who sat with them. Their recommendations were then "translated into action by the Air Staff and pertinent field commands through the coordination of the [SAB] Military Secretary."[6] After the Air Force and MIT, acting on ADSEC and SAB recommendations, created the Lincoln Laboratory there was no further need for the committee and on Dr. Valley's recommendation, the SAB formally dissolved it in January 1952.

ADSEC's role in strengthening the nation's air defenses was a very significant one and its members showed the way "to many promising developments in this field," General Vandenberg wrote Dr. Valley. General Twining later added his compliments, noting he was "satisfied that our best minds are [now] working hard on air defense and I think the board can well be proud of its activities."[7]

Meanwhile, President Truman had committed American military forces to stopping Communist aggression in Korea. Whereas the Russian atomic test had stimulated considerable concern among U.S. military departments, the nation as a whole had failed to appreciate the Soviet ability to master modern sciences and engineering. As a result, American military activity in the months immediately following that event continued to steadily decline. This trend was abruptly reversed when the North Korean Communist regime attacked the Seoul government. Congress promptly authorized a massive budgetary increase for military research and development as well as for operations. With its new Air Research and Development Command still in a formative stage, the Air Force turned to the SAB for help in allocating wisely its share of this increase. As a result, most of the SAB activity during the next year, as one report noted, "was pointed directly toward supporting the augmented USAF Research and Development program resulting from our commitment in Korea and the increased probability that we would be committed elsewhere."[8]

SAB's first important undertaking following the outbreak

See notes on page 183.

of war entailed a review of the Air Force guided missile program. Lt. Gen. Samuel E. Anderson, then Director of Plans and Operations on the Air Staff, asked the SAB on July 17, 1950, to recommend changes in emphasis in that program. Specifically, he asked the board to indicate which guided missiles under development showed most promise and when the Air Force might succeed in putting them into operation. He also asked for SAB's opinion on how the program could be expedited if additional funds were allocated.[9]

Dr. Ridenour accepted chairmanship of the SAB ad hoc committee formed to handle the request, and Drs. Astin, Francis H. Clauser, Donovan, and Stever accepted membership. Joined by officers from the Air Staff and from other Air Force agencies engaged in various aspects of the program, the committee met in Santa Monica, Calif., on July 25-29, received thorough briefings from contractors, and framed recommendations. Decision on the matter was deemed sufficiently urgent for the Air Staff representatives to report the committee's findings to the Pentagon "immediately upon formulation." The committee's final views, recorded and submitted on the last day of the meeting, coincided for the most part with those of the Air Staff members in attendance and actions were undertaken at once. In a summary statement, the committee noted that in the past there had "been a tendency to regard guided missiles, as such, as being special items demanding special treatment." The committee did not agree with this point of view, urging the Air Force "to regard guided missiles . . . as being the natural next steps in the cultivation of various aspects of air warfare" and so hasten their entry into the operational inventory.[10]

So many requests for SAB assistance had piled up in the two months following the outbreak of the Korean war that Dr. von Karman called the board into session for a full week beginning September 11, 1950. A major agenda item was to assess the planned allocation of the increased research and development funds. Soon after war began, the Secretary of Defense had directed the military services to forecast their needs for the next decade, to include funds they needed immediately (in supplemental fiscal year 1951 appropriations) to expedite key projects that had lagged for lack of financing.[11] Though General Saville, Deputy Chief of Staff/Development, drafted the Air Force reply by early September 1950, he informed his

See notes on page 183.

superiors (in one of the most gratifying compliments thus far extended the SAB) that the report had to be "considered tentative" until the board had reviewed it.[12]   Following his presentation, the board divided into four functional committees—air defense, strategic air, tactical air, and air transport—and, for the next two days, studied the issues and contributed written recommendations.   Afterward, the board regrouped into its regular panels to discuss and record their findings on problems within their technical specialties.   Since panel members had sat on the functional committees, they had the requisite background to consider technical programs in the light of system requirements.[13]   Dr. von Karman forwarded the committee and panel reports to General Vandenberg on September 19, 1950, at the same time offering his personal summation of the factors "upon which the successful implementation of the augmented R&D program will heavily depend."[14]   The final consensus was that the board's approach at the meeting was an excellent one and its "efforts most useful."[15]   For his part, General Saville, speaking of the findings given in the four committee reports, said: "I would like to say on behalf of the Chief of Staff that if you don't do another thing [in follow-up to the meeting] we sure appreciate it, because these four reports show evidence of the soundness of [your] process and we get information and guidance out of this, gentlemen, that we can get no other way."[16]

The increase in funds now put the Air Force in a position to proceed from the planning to the operational stage with its new Air Research and Development Command.   As the ARDC commander, Maj. Gen. David M. Schlatter, told the board at their September 1950 meeting, he had been "in command of a piece of paper" until recently.   Now, with the help of SAB's Dr. Mervin J. Kelly and Dr. Donald A. Quarles, he had begun the job of building ARDC into an operating command.   As he summarized their labors:[17]

> We proceeded from a thorough study of the philosophy of the concept of [the Ridenour] report, the Air University report, various correlated documents and studies. [We made] a thorough physical study of the existing plant facilities, went briefly through most of their programs, [and] took a look at some other organizational structures, such as the Bureau of Ordnance, the Bell Laboratory . . . and Union Carbide and Chemical. We have proposed a plan for making the command operational, and I expect the Chief of Staff to act on that very soon. This planning period that we have gone through has been, I might say, deliberately deliberate. We wanted to make

See notes on page 183.

sure that we were going to find the best answer for the Air Force
... to meet the philosophy and concept of [the Ridenour] report.

As General Schlatter had predicted, General Vandenberg
quickly approved the plan for building up ARDC, directing
on October 12, 1950, that ARDC begin functioning "as an independent, self-sufficient major Air Force Command" by May
1951. In the interim, AMC would continue to expand the Air
Force's technical competence to meet the increasing requirements of research and development, production, maintenance,
and supply engineering.[18] Creating openings for scientists and
engineers and filling the positions were two different problems,
of course, and General Twining, Vice Chief of Staff, asked Dr.
von Karman to appoint a special SAB working group to explore
ways—whereby the Air Force might attract people in universities and civilian laboratories to help with the current rearma-
ment effort. The group was also invited to study and comment
on any other aspect of Air Force research and development.

Dr. Ridenour accepted von Karman's invitation to chair
the working group, picked about 20 members (from the SAB,
government, industry, and universities), and convened meetings in late January and early February 1951. Here they drew
up a recommended program for an emergency augmentation of
the USAF research and development staffs and assigned individual members of their group specific responsibilities for helping the Air Staff to implement the program. In a recommendation destined to have even greater significance in subsequent years, they urged the Air Force to create more weapon
system laboratories in order to properly integrate the "rich
variety" of new developments achieved in components since the
end of World War II.[19]

The working group met often through the summer of 1951
then gradually phased out. While producing no formal final
report, it did generate a great number of important projects
during its brief lifetime. Among these were (1) Vista, a Caltech study of airground tactical warfare, (2) Buffalo Bill, a
Cornell Aeronautical Laboratories study for modifying tactical
experimental aircraft, weapons, and equipment preparatory to
arranging for their development trial, and (3) the Harvard
Logistics Study on air transport and general Air Force logistics
problems. As Dr. Doolittle later assessed the working group's
overall accomplishment, it "provided the initiative and was of
direct influence in the formulation of [activities] which have

See notes on page 183.

had a great effect on our national military posture and structure."[20]

Fortunately, the nature of the Korean War enabled the Air Force to apply various expedients by which it eventually surmounted the grievous deficiencies in aircraft armament which plagued its early operations there. To help insure that the Air Force would not be outgunned in future conflicts, whatever their nature, General Vandenberg asked Dr. von Karman in May 1951 to perform a study of armament requirements.[21] Dr. John A. Hutcheson (who soon after became a SAB member) agreed to chair the SAB ad hoc committee formed to do the study. Joining him were SAB members Draper, Kent, Charles C. Lauritsen, and Louis T. E. Thompson, all from the Explosives and Armament Panel, and invited members Mr. H. T. Hokanson and Mr. Edgar Schmued (who later joined the board)—all, as one report noted, "outstanding individuals with broad experience in all phases of USAF armament activities."[22] Beginning their work the week of June 18, 1951, they submitted an interim report on June 30 and the final report on October 4.[23] The major obstacle to an effective armament program to date, they felt, was the shortage of technically qualified personnel. Earlier SAB and Air Staff studies had pointed up this problem and the Hutcheson Committee noted their "profound disappointment" that action on the recommendations of these studies had lagged. Consequently, they again urged the Air Force to broaden its armament competency through such means as offering direct commissions to outstanding persons in this field, expanding the Air Force Institute of Technology's armament curriculum, and affording officers greater opportunity for graduate training in civilian institutions.[24] In the years following, the SAB was gratified to observe the incorporation of these and other Hutcheson Committee proposals into the Air Force armament research and development program.

The increased demand for SAB services induced by the Korean war resulted in an increase in SAB membership from 46 in March 1950 to 62 by June 1953. Five of the new members staffed the one new panel formed during these years. This was the Physical Sciences Panel, approved by the Executive Committee in May 1951, formed in July of that year, and con-

See notes on page 183.

vened for the first time the following October at AFCRL. Dr. George B. Kistiakowsky accepted chairmanship and Drs. Baker, Edwin H. Land, Overhage, and Milton S. Plesset membership. Collectively expert in such areas as metallurgy, optics, and solid state physics, the new panel brought added strength to the board in the field of basic physical scientific research, with particular emphasis on optics, photographic processes, and other techniques required for conducting aerial reconnaissance missions.[25]

Two of the new members, Dr. Doolittle and Dr. Kelly, both of whom joined the board soon after the Korean War began, accepted appointment as co-vice chairmen in late 1950, replacing Dr. Dryden in that position when the press of other duties forced him to take a less active role in board affairs. To accommodate the other new members and, at the same time, keep panels at a workable size, SAB officers created a new "members-at-large" category. Most of the persons assigned here were in government employ and considered to possess sufficiently broad experience to operate effectively in whatever capacity the SAB might require their services.

Membership policies remained the same throughout the Korean War period except for a revision in appointment procedures. Because the Air Force had been extremely slow in concluding contract actions, the Executive Committee obtained permission to change board appointments from the fiscal to calendar year. This enabled the Air Force agency which processed these matters to act on board appointments at a time when it was not swamped with other end-of-the-fiscal-year contracts. As a result, board members appointed for fiscal year 1951 automatically served an additional six months, through December 1952. From that time, appointments were for a calendar year.[26]

The top position in the SAB secretariat changed from military to civilian in late 1950 when Colonel Walkowicz became executive to the Air Force Chief Scientist and Mr. Driscoll replaced him.* Mr. Chester N. Hasert accepted the position upon Driscoll's resignation from government service in

---

*See Appendix B for a roster of Air Force Chief Scientists. The position was created on SAB recommendation and first occupied by Dr. Ridenour in September 1950.

See notes on page 183.

1952.* Meanwhile, Majors Mark P. Maier and Daniel D. Whitcraft had joined the staff as assistant secretaries.[4]

Two important changes in board policy and procedure occurred during the Korean War, and both were included in a June 1951 amendment of the SAB regulation (AFR 20-30). The first implemented an October 1950 instruction from General Vandenberg making the Deputy Chief of Staff/Development responsible for monitoring the implementation of SAB recommendations and for apprising the Chief of Staff and the SAB chairman on the status of those recommendations. As discussed earlier, Vandenberg had levied this requirement on each deputy in 1949; the change strengthened the procedure by centralizing responsibility for it in logical hands.[27]

The second change concerned the appointment of the SAB military director, a position automatically held since 1947 by the Director of Research and Development. Now it was decided that the Chief of Staff would appoint to the position the general officer on his staff most suited by education and experience to perform the duty. As it worked out, the change eventually resulted in raising the position from directorate to deputy level on the Air Staff; from 1952, the Deputy Chief of Staff/Development served as military director without exception.††

---

*Mr. Hasert had served as a military staff member on the wartime Scientific Advisory Group.

†The top position in the secretariat held the title of "executive secretary" for several years. Subordinate positions carried various titles. Since their duties remained constant despite the name changes, they are referred to as "secretary" and "assistant secretary" throughout the paper.

††The position was renamed DCS/Research and Technology in 1961 and became DCS/Research and Development in 1963.

See notes on page 183.

CHAPTER FIVE

FOR A MORE SUCCESSFUL MARRIAGE

> We are here to determine how we can improve the SAB. I am trying to see that science and the Air Force have a successful marriage. As I see it, the SAB has three chores: (1) Try and answer some of the day to day problems of the ARDC and DCS/D; (2) try and answer questions of the Chief of Staff and, of course, the Secretary; and (3) conceive new and future thoughts. The board must not become so involved in the day to day operations that they do not do the other two jobs.
>
> ——James H. Doolittle*

With the waning of the Korean War, the SAB found time to correct weaknesses that had appeared during the war in the board's panel structure. Some members felt that the basic board composition ought to be realigned in a different fashion so that members whose panels had not been very active in recent months might be better employed. But Dr. von Karman quashed further consideration of this at a March 1953 Executive Committee meeting with the statement that he considered the present philosophy of board organization quite appropriate.[1] The committeemen then concurred in actions that subsequently resulted in the formation of two new panels and the dissolution of two old ones.

One new panel, called Nuclear Weapons, took cognizance of the impact the thermonuclear breakthrough portended for

---

*In remarks at SAB Executive Committee Meeting of June 15, 1954.
See notes on page 184.

Air Force operations. In the past, the Explosives and Armament Panel had handled most of the board's nuclear weapon investigations. However, the successful Eniwetok Island "Mike" shot in October 1952 opened vast new areas of possibility for the military use of these weapons and the new panel expanded SAB expertise in the science.* The panel took as its charter all aspects of nuclear development except propulsion (which remained with the Fuels and Propulsion Panel). It also acted as a link with the development laboratories of the Atomic Energy Commission to insure "effective exchange of information in both directions." Dr. John von Neumann, whom Dr. von Karman dubbed the "midwife at the birth of the Panel," accepted chairmanship with Dr. Hans A. Bethe, Dr. Norris E. Bradbury, Prof. David T. Griggs, Dr. L. Eugene Root, Dr. Herbert S. Scoville, and Dr. Edward Teller members.[2]

A second change introduced at this time eventually transformed the Physical Sciences Panel into the Reconnaissance Panel. As noted earlier, the former had been created in mid-1951 presumably to concentrate on such Air Force reconnaissance matters as optics and photography. However, as chairman Dr. Kistiakowsky explained, the panel's time had been totally consumed in giving policy guidance on the creation and build up of ARDC's Office of Scientific Research (OSR). Now that OSR was in operation, with its own scientific advisory group, SAB officials had to decide what to do with the panel. Temporarily, they dissolved it, appointing Dr. Kistiakowsky to the Explosives and Armament Panel and carrying its other members (Drs. Baker, Land, Overhage, and M. S. Plesset) as members-at-large.[3]

Meanwhile, two top-level study groups—Vista and Beacon Hill—had pointed up the many severe problems currently facing the Air Force in intelligence and reconnaissance. Consequently, SAB officers decided that the impressive talent released by the dissolution of the Physical Sciences Panel could best be employed in these areas and formed the Intelligence Systems Panel, which met for the first time in August 1953. In addition to the members of the former panel, Dr. Donovan, Dr. Duncan E. Macdonald, Mr. Stewart E. Miller, and Mr. Phillip G. Strong subsequently accepted membership, under Dr. Baker's

---

*Impetus for creation of the Nuclear Panel stemmed from a January 15, 1953, letter from Dr. Doolittle to Dr. von Karman in which Doolittle, after tracing recent thermonuclear advances, noted it was his feeling that "we cannot afford to delay the formation of such a panel."

See notes on page 184.

chairmanship.[4] In 1955, Dr. Overhage, who succeeded Baker as chairman, changed the name to the Reconnaissance Panel.[5] The "intelligence label" was discarded, Overhage explained, because the panel did not deal with what was commonly called by that name. Its members were interested in the physical techniques of collecting and processing certain types of data which ultimately contributed to the development of a pattern of information called intelligence, he said. However, the final result involved many steps which, as a group, the panel was "neither competent nor eligible to discuss."[6]

A final panel adjustment in mid-1953 transferred the duties and most of the members of the Guided Missiles Panel to the Aircraft Panel. The board had considered merging the two panels in 1951, but decided against it at that time because the Air Force still handled the subjects under separate organizational compartments.[7] Since then, however, the Air Force had concluded that flight had to be treated as an entity, whether it took place in the atmosphere or in space.[8] All Aircraft Panel members remained after the change except Dr. Root, its chairman, whose talents were required on the new Nuclear Weapons Panel. Dr. C. B. Millikan accepted appointment as Root's successor. All members of the dissolved Guided Missiles Panel transferred to the Aircraft Panel except Dr. Stever, who joined the Explosives and Armament Panel.

Having adjusted the panel structure, SAB undertook a review of other procedural aspects which appeared to need strengthening. First, however, the board underwent an unofficial but quite real change of leadership. Since the summer of 1950, when he undertook a study of the state of aeronautical science in the member nations of the North Atlantic Treaty Organization (NATO) for General Vandenberg, Dr. von Karman had been fighting a losing battle to give an equitable share of his time and energy to both the SAB and the new project. On his return from Europe, von Karman had suggested that the Air Force "explore . . . this problem of using European science for common defense."[9] General Vandenberg had passed the proposal through higher echelons to the Standing Group for NATO, who liked it and had the U.S. Air Force appointed Executive Agent for organizing a conference of European scientists to consider the matter. Dr. von Karman and the SAB secretariat handled the details of getting the activity in motion. The scientists met in the Pentagon in February 1951, after which

See notes on page 184.

the Joint Chiefs of Staff appointed an inter-service committee to act on the scientists' recommendation to form a NATO group for coordinating aeronautical research and development. In January 1952, the committee approved establishment of the Advisory Group for Aeronautical Research and Development (AGARD) and, the next month, Dr. von Karman accepted General Vandenberg's invitation to be its first chairman.[10]

Dr. von Karman had chaired the fall 1952 and spring 1953 SAB general meetings. But from this point through the remainder of his SAB chairmanship he found it necessary to rely on other SAB officers to bear the weight of matters. As the SAB secretary, Mr. Hasert, noted during a visit with him at his AGARD offices in Paris during this period, "the boss [was] very tired" and it was obvious, that in the future he could not be expected to contribute very much of his energy to SAB affairs.[11] Fortunately, Dr. Doolittle, board member since 1950 and one of the two vice chairmen, was able to shoulder the load.*

One of Dr. Doolittle's first acts was to sponsor a reduction in board membership. As noted earlier, membership rose to over 60 during the Korean War. The Executive Committee first expressed concern with the matter in late 1952, when, as Dr. Root expressed it, they concluded that the "number of members [ought] to be kept to a total which permits a continuing high degree of board flexibility and effect."[12] Dr. Doolittle agreed, noting that the board had been gradually "getting a little large and a little hard to handle." He felt a smaller board would be more flexible. He also felt the Air Force should not tie up too large a percentage of the nation's top scientific talent, explaining that "the Army, Navy, and other agencies also need good people and some of these people should have the opportunity to tie in with the other services; if we have a small board, we can have more time from the individual members of the board [and] when they rotate they can then serve another service."[13]

The Executive Committee concurred and, in October 1953, voted to restrict the size of the board to 50 members. At the same time, it authorized the panels and special committees to

---

*Dr. Doolittle accepted the unique post of Acting Chairman on November 25, 1952, in compliance with Dr. von Karman's request of this date that he "act for me in those instances when I am out of the country . . . when it is necessary to have on-the-spot decisions regarding plans and management of the SAB."

See notes on page 184.

use consultants for limited periods to help handle any special projects that arose. Panels and special committees had done this on occasion during the Korean War with gratifying results. In answering a member's query on the difference between a member and a consultant, Doolittle suggested that the following serve as a rule-of-thumb definition: "These consultants . . . do not come to the full board meetings unless they are specifically invited. If the chairman [of a panel or special committee] wants them invited, he can recommend this to the secretariat who will then decide whether [they are] to be invited."[14] Once asked to serve on a panel or committee, consultants would receive the same payments (if they so desired) and courtesies as regular members for the duration of that association. In subsequent years, successful service as a consultant came to be viewed as one of the criteria for an individual's suitability for board membership. Too, persons who found they did not have the time to continue regular membership were usually willing to serve as consultants when their specific talents were in demand. Thus, in accordance with Dr. Doolittle's concept of the role, there was little difference between members and consultants in terms of value to the board and the weight given their participation in board affairs.

Having placed a ceiling on board membership, Dr. Doolittle now suggested that they institute a system for membership rotation. Since the board's founding, members had accepted membership for one year then renewed annually for as long as the SAB needed them and for as long as they had time to serve. The danger here was that the SAB might come to rely so heavily on certain individuals it would fail to acquire the "new blood" so essential to its long-term health. The subject received Executive Committee study until the summer of 1954 when Doolittle translated the various suggestions into a written policy statement which General Twining approved on August 2, 1954.[15] The policy, placed in effect in 1955, stipulated that current members who wished to remain on the board would receive appointments ranging from one to three years. New members joining in 1955, and from that time on, would be appointed initially for one year. At the end of that time, all parties being in agreement, the board would offer them a "regular" three-year term. Once a person dropped from membership, a year had to elapse before he was again eligible to serve. In other words, while the policy did not restrict total time of membership, it did limit

See notes on page 184.

any one "hitch" to a maximum of four years.[16] SAB officers carefully stipulated that the policy was an internal matter whose terms were not to be publicized. This was essential, as Dr. Soderberg pointed out, in order not to tie the hands of future board officers who might wish to depart from its strictures.[17] As noted later, and as the membership rosters in the appendix reveal, the board never managed to adhere very closely to this policy and it eventually died a quiet death. At the same time, the basic problem of freeing the board of old-timers in order to make way for the young remained a point of recurring concern.

In a final adjustment of board mechanism, Dr. Doolittle eliminated the members-at-large category from the board roster. During the Korean War, it will be recalled, the board had extended such membership to several persons, most of whom were in government employ. From seven in 1951, this group increased to 15 over the next two years. In Doolittle's opinion, "the members-at-large were overtaking the members . . . getting completely out of hand." On his recommendation, the Executive Committee dropped the category, reassigning some of the people to ex-officio status and the remainder to panels. The ex-officio members, Doolittle explained, "would not be regular members, but you could call upon them for a particular job."[18]

Since members had occasionally questioned the value of certain general board meeting procedures over the years, Dr. Doolittle asked the Executive Committee in the summer of 1953 for an opinion on whether these meetings should continue and, if so, how they might best proceed. The committee agreed that they were necessary and recommended they continue to be held at least twice a year—in the spring and in the fall or early winter.[19] They also approved continuance of a practice adopted during the Korean War of building each meeting around a subject of major and current importance to the Air Force, then holding it at an Air Force installation concerned with that subject. Through 1951, all board meetings had convened in the Pentagon with occasional side trips to Wright Field. While field commanders had joined them there and presented their problems, the members had not been able to actually inspect many of the projects or weapons discussed. However, the two 1952 and spring 1953 meetings were held at field command headquarters, a practice the Executive Committee now elected to continue.

See notes on page 184.

Finally, the committeemen adopted Dr. Doolittle's proposal to devote future meetings "primarily to informational briefings and . . . inter-panel exchange of ideas, but without the expectation of necessarily solving problems on this occasion, or of producing reports containing formal recommendations."[20] This was a significant change in procedure. Before 1953, the board had always submitted formal recommendations to the Chief of Staff on problems discussed during each general meeting. Many members had voiced criticism of the practice, however, feeling they could not do a reliable job in such a short time and with the frequently incomplete data presented them. In future years, the board prepared and issued reports after each meeting, but restricted them to an account of the briefings given and of any significant discussions which took place during panel or plenary sessions. In short, the reports ceased to be official recommendations requiring official responses. When board officers felt an agenda topic required further consideration, they assigned it to a panel or special group who then gave it requisite study and submitted formal recommendations.

Questions on the value of the general board meetings continued over the years, but board officers always decided that the criticisms reflected only a minority view. Dr. Doolittle noted to General Twining in 1954 that the Executive Committee favored continuing them because

> we feel that these meetings have been quite successful in familiarizing our members with the Air Force's operational problems to which their ideas, in the final analysis, are aimed. In addition, the SAB has strengthened its ties with the people who are ultimately responsible for applying the products of technology to the requirements of operations. Moreover, these meetings have been useful in introducing the SAB to our operational people as a unique and highly valuable Air Force resource in the exploitation of technological opportunities as well as the solution of broad technological problems.*[21]

A criticism by Mr. Trevor Gardner, Assistant Secretary of the Air Force for Research and Development, focused attention on a final major problem facing Dr. Doolittle and the Executive Committee at the end of the Korean War. Following the spring

---

*Nearly 10 years later, the Executive Committee continued to explain and defend the general board meetings in about the same words. While little or no real results had appeared to stem from full board meetings in recent years, the minutes of a December 1963 Executive Committee meeting reported, it had to be recognized that "the main purpose of these meetings is education of the members and that the most benefit accrues from the 'give and take' discussions following the briefings."

See notes on page 184.

1954 general meeting, Gardner said, several SAB members "made comments to the effect that the SAB is not a useful body for the Chief of Staff and should be concerned with problems more directly related to ARDC."[22] None of the SAB officers denied the validity of the last part of the criticism, having already discussed the matter at a March 1954 meeting. Much of the problem, they felt, stemmed from the fact that SAB reports went to the Air Staff for action with ARDC receiving only "information" copies and, as Dr. Doolittle observed, these "sometimes got lost." In short, ARDC sometimes remained unaware of SAB recommendations or was not called on for action. As the board had learned when research and development was still a function of AMC, it had to work closely with both the Air Staff and the field command if it were to succeed in having its recommendations acted on.[23]

At first, the board officers considered altering the procedure for dispatching SAB studies so that copies (after their approval by the Chief of Staff) would go out simultaneously to the Air Staff and to ARDC for action. But this was opposed on the grounds, as one member expressed it, that "if we attempt to make action copies for [ARDC] it is possible for us to contradict actions from the Air Staff." As finally resolved, the Deputy Chief of Staff/Development assumed responsibility for dispatching copies of SAB studies simultaneously to his directorates (and other appropriate Air Staff agencies) for comments on which to base the official Air Staff reply and to ARDC "for comment or advance information." He also assumed responsibility for preparing "action" correspondence to ARDC (or other appropriate command). This revision and strengthening of the procedure for processing SAB recommendations (issued initially as Deputy Chief of Staff/Development Office Instruction 11-9, April 9, 1954) eventually emerged as a Headquarters USAF Operating Instruction (HOI 80-7).* Essentially, as Dr. Doolittle pointed out, it formalized for really the first time how SAB reports were to be handled at the working level. In short, it brought a long-overdue orderliness to the procedure of handling SAB recommendations and assured ARDC of receiving them promptly.[24] However, as subsequent chapters relate, it did little to quiet criticism that SAB talents were not being made sufficiently available to ARDC.

---

*HOI 80-7 appeared initially on July 8, 1960, and subsequently underwent three revisions: November 12, 1962, July 8, 1963, and April 8, 1964.

See notes on page 184.

Concerning the first part of the criticism voiced to Mr. Gardner, that the SAB "was not a useful body," Executive Committee members assured him that they did not share this view. _For example, Dr. Ivan A. Getting, a charter member of the board, noted that there had been times when he might have agreed with such a criticism, but he was now convinced that "a lot of good has come out of the SAB." Dr. Doolittle agreed, closing the long hours he and his fellow SAB officers had devoted during 1953-1954 to charting a meaningful course for the board in the post-Korean War era with the recommendation that it should proceed on the assumptions that (1) it was an essential organization and could be made even more so through greater liaison between panels, ARDC, and the Air Staff, and (2) individuals on the board should not wait for directives but "be on the alert [to] conceive fundamentally new ideas [and] concepts and . . . give [them] to the Air Force."[25]

See notes on page 184.

CHAPTER SIX

NEW LOOK AT THE FUTURE

> The development of thermo-nuclear
> devices; tremendous progress in
> electronics and power plants; rou-
> tine operations at Mach numbers
> which, only a few years ago, were
> considered impossible of attain-
> ment; and . . . the mounting prob-
> lems of defense against a ruthless
> enemy with similar capabilities—
> all these . . . make it mandatory
> to take stock . . . so that we know
> in what direction to go or not to
> go.
>
> ——Donald L. Putt*

During the June 1954 Executive Committee discussion of board tasks, Dr. Wexler commented that the one which called for the SAB to stay ahead of the Air Force in conceiving new ideas and concepts was the most pleasant to perform. Dr. Doolittle agreed, adding however that it was also the most difficult. Precisely how difficult was being demonstrated by discussions and study then under way within the board on the advisability of attempting to bring *Toward New Horizons* up to date.

In the years since the publication of that study many persons had expressed hope that the SAB would update it. The board had given the subject considerable thought in early 1949 after General Vandenberg noted informally that he would like to see the board undertake a lesser-classified revision of the original which could be widely circulated among USAF field grade officers.[1] However, the Russian atomic test and the Korean War caused the board to go in other directions.

---

*In address to the SAB, October 19, 1953.
See notes on page 184.

At the end of the war, the subject inevitably came up again, broached this time by General Putt, then serving as the ARDC commander. "We in the military feel that such significant strides have been made in the basic sciences during the past ten years that another look is required at the various technical fields which contribute to modern air power and provide the foundation upon which it is built," Putt told the SAB at their fall 1953 meeting. He promised ARDC assistance and predicted that an updated version "would rank in impact alongside [the original] and the Ridenour Report and would, like these, contribute to Air Power the element of indisputable superiority which is so vital to the preservation of our nation."[2]

Soon after, Putt expanded on his views to General Craigie, SAB military director. "What we need," he said, "is an intensive long-term look at the trend capabilities of these technical areas which will contribute most to the development of Air Force equipment in the next ten years." A new SAB study which provided "foreseeable trends" in the applied and basic research areas of each technical field of interest would give ARDC a "firmer foundation on which to base our immediately pressing, broad planning and management decisions." It would also point out "the direction our supporting facilities, equipment, manpower, and budget requirements must take for the next decade or so."[3]

In short, Dr. Doolittle observed, Putt wanted "to have the old subjects modernized, add new ones and new concepts that might not have been thought of at that time, [and] . . . show places to go and how to remove bottlenecks that are holding us up."[4] Even after simplification into Doolittle's working vernacular, the request still constituted the largest task asked of SAB since its World War II days. The question now was, should SAB undertake it and, if so, how should it proceed?

As an initial step, Mr. Hasert discussed the subject with Dr. von Karman in Paris, who gave it "a very cautious and somewhat reserved" reception.[5] However, to allow a full airing of views, von Karman joined the panel chairmen at a special meeting in the Pentagon during January 1954. They quickly agreed that their role as a part-time advisory board would not permit the production of a set of volumes comparable to *Toward New Horizons*. To see what they could do, von Karman asked each panel chairman to survey his "field of cognizance" and, for the next board meeting, report briefly on those areas which

See notes on pages 184 and 185.

showed the most promise for long-range planning. The reports were to be brief, representing simply the "informal, undocumented raw ideas by experts in the specific fields."[6]

Accordingly, panels met and recorded their views which panel chairmen then presented at the March 1954 meeting. Dr. Getting's preface to the Electronics and Communications Panel report aptly described the intent and nature of them all: his panel had "tried to confine . . . attention to the strict problem laid before it, namely, sort of a dry run on what kind of prediction we could make on the spur of the moment for the next ten years."[7] Consequently, the reports fell far short of being reliable harbingers of things to come, their authors carefully pointed out. At the same time, they did pinpoint panel impressions on many significant Air Force matters and, in many cases, revealed the major areas of panel interest over the past years.

Dr. Hastings reported that his Aeromedical Research Panel "had serious doubts about the ability of itself or others to be able to put down in black and white in 1954 adequate guesses or forecasts about the future that would be meaningful." This was especially so, he noted wryly, since "the central preoccupation of our panel, homo sapiens, has not undergone any observable changes, unfortunately, in either his psyche or soma for at least four or five thousand years." Biological scientists had developed many new techniques, however, and compiled new data of a number of types since the end of World War II. For example, they now had a far better understanding of how the body adjusted to heat, cold, and emotional stress. They felt this new knowledge might enable the Air Force to reinvestigate "some of the old problems that we have had with us a long time . . . with some better probability of success." Another problem that aeromedicine faced in the future was the challenges that space operation posed. Weightlessness was one such challenge. "We could recommend that intensive research be done on this problem," Hastings said, "but at the present time there does not appear to be any satisfactory method by which to approach it short of building a satellite, orbiting it about the earth and seeing what happens either to human volunteers or to experimental animals." Of the future, generally, his panel could "only foresee increasingly difficult problems with respect to trying to get the human body to adjust to these new environments."[8]

See notes on page 185.

Dr. Millikan's Aircraft Panel believed that the most important and vital subject in aerodynamics for the next 10 years was "the field of hypersonic flows and, in particular, hypersonic flows with stagnation temperatures which may run up to the order of thousands of degrees." In the research vehicle program, the panel felt that the Air Force should promptly initiate two projects—one in unmanned rockets for hypersonic speeds and another "involving manned aircraft to reach something of the order of Mach Number 5 and altitudes of the order of 200,000 to 500,000 feet." The members also thought that the Air Force should develop three radically new types of aircraft over the next 10 years if it could afford the large development costs. These were the vertical take-off and landing combat aircraft, the rotary-wing tranpsort and assault aircraft, and the nuclear-powered aircraft augmented by chemical-fuel boost.[9]

Dr. Getting's panel agreed that electronics "had reached a point where a reasonable promise can be made in meeting a specific operational need." The development of reliable and smaller components—transistors, ferrites, magnetic storage, and amplifying devices—was perhaps the greatest achievement in electronics since World War II. This had opened the door to various "startlingly new possibilities," particularly in data handling and analysis. Concerning the direction electronics research and development should take in the future, panel members believed this would be mostly determined by the estimates put forth on the nature of future wars. Consequently, they recommended that an analysis be made as soon as possible "to fix the scope of warfare and establish the framework from which the direction of electronic developments should proceed." The current tendency to rely completely on atomic warfare, they warned, "results in a development of electronic equipment peculiar to that type of warfare and it is not clear that such equipment will provide the adequate arsenal for the United States."[10]

Dr. Kent's Explosives and Armament Panel believed the Air Force had to improve fire control systems if it hoped to exact the fullest potential from future new weapons and explosives. Several panel members expressed their view that "if as much effort were used on reducing the miss distance as has been used on making bigger and better bangs, we would very soon have very much improved fire control." On Kent's request,

---

See notes on page 185.

Dr. Draper, the panel's foremost expert on fire control, pursued the point in detail, noting that

> we have had many periods where our eyes have been directed too much in the direction of making bigger and bigger bursts and have felt that fire control was not capable of doing very much in the first place, and in the second place, that we didn't need to be very good anyway because guided missiles and big bangs were going to take the place of it. I have a feeling that with the techniques we now have pretty well in hand, and with the components which are now pretty well available, one should be able to make lightweight and fast fire control gear which should do the best possible job toward laying the centers of impact close to the target. Now I am quite aware that the whole job can't be done with the fire control system because actually the main problem is to get the information about the target early enough so that you can get ready to do the job. But once having that information, which of necessity must come from the electronics end of the game, it seems to me that the obligation is on the fire control to do the best job in the very shortest possible time. I think this is a subject that has not been given as much attention in the past as it can be given, and I predict that we are going to be able to get quite a bit of information in this direction.[11]

Mr. Addison M. Rothrock, acting for chairman Soderberg, reported that the Fuels and Propulsion Panel agreed with the current USAF supersonic turbojet engine development program. They further proposed that the Air Force begin exploratory studies "directed towards finding ways of improving efficiency in low altitude subsonic flight while retaining efficient subsonic and supersonic performance at altitude."[12]

Dr. Wexler pointed out that the atmosphere remained the chief subject of the Geophysical Research Panel: "As long as we send things [there] we must learn to become acquainted with and live intelligently with [its] peculiar properties," he said. In meteorology, progress had been and would continue to be "agonizingly slow," and Dr. Wexler doubted if there would be any really revolutionary advances in the next 10 years. At the same time, he saw continuing improvements in areas which were enabling today's meteorologists to predict weather changes "in scope and detail undreamed of a generation ago." In summary, his panel believed that "the brightest spots on new horizons for geophysics . . . [were] in increased use of electronics in probing the atmosphere, in acquiring truly global coverage of weather observations and solar and cosmic data above the atmosphere, and in streamlining and rendering more automatic the procurement and processing of weather data for forecasting and climatological summaries."[11] All of these yielded "an

See notes on page 185.

improved understanding of the atmosphere as a necessary step in large-scale weather control."[13]

Dr. von Neumann spoke extemporaneously from notes he had taken at his Nuclear Weapons Panel meeting. At the time *Toward New Horizons* was written, he pointed out, very little was known about nuclear weapons and even this was known by just a few people. Also, "the main approach to the subject at that time represented a single type of bomb." Since then, he said,

> the knowledge is more diffused and should be diffused still further, since it concerns considerably more people, and the variety of weapons types and underlying principles has increased a very great deal. In addition to the extensive technological development which took place during these eight years, there has also been a complete change in the underlying economic-political-strategic position: nuclear weapons are no longer expensive, they are no longer scarce, and they are no longer a monopoly of the U.S. These things are known, but they still need to be repeated—I do not think that our thinking has assimilated them to the extent to which it should.
>
> I would like to reemphasize that nuclear weapons are no longer scarce. It is much easier to increase our capability in this area by any factor any one may require than to increase correspondingly those capabilities which are needed conjointly with it, in order to deliver the weapons. In other words, one must no longer consider the nuclear component as the hardest part of the problems involved in the weapons systems of which they form part—they are now among the least difficult and most flexible parts of such systems.
>
> . . . it is not at all clear how far the invasion of other fields in weaponeering by the nuclear weapons will go. This question is really a fundamental one, and one should view it in its proper proportions. Thus, it is quite possible that nuclear weapons might replace all artillery, or . . . they may replace 90 percent or more of conventional high explosives' production, and so on. Questions of this type now call for serious consideration. On the other hand, one must also think about the possibility of limitations in the use of nuclear weapons which are not military at all, but which are due to the fact that the use of nuclear weapons has consequences far beyond their strictly military effects. There exists in this regard a great deal of romanticism—mostly with an inverted sign—but this does not alter the fact that these questions must be considered seriously and realistically. In particular, we will have to give a great deal of attention to the alternatives of the strategic use of nuclear weapons, the more so because this strategic use has been dominating most of the past thinking on this subject.

The uncomplicated conclusion, Dr. von Neumann said, was "that in nuclear weapons we have something in our hands which has very broad consequences and applications."[14]

See notes on page 185.

Dr. Baker's Reconnaissance Panel stressed the need for adequate funding of intelligence in the next decade for developing "aerial reconnaissance systems capable of gathering information regardless of enemy counteraction."* This called for "special reconnaissance vehicles not encumbered by being designed for other duties at the same time." While the Air Force could not risk disclosing secret devices by sending atomic bombers over enemy territory in peacetime, it could conceivably perform the operation with "a high altitude reconnaissance airplane having no other function." The panel believed the Air Force should "at least . . . have such vehicles ready and on call." Concerning the intercontinental ballistic missile threat, Dr. Baker noted that "our task in intelligence should be to find the launching sites and supporting installations, to keep these under surveillance, and to anticipate actual hostilities by a sufficient margin to prepare our own countermeasures." As for the immediate task facing the panel, the members believed they should "consider our most important function a dynamic one of seeing that immediately valuable projects are kept moving along without important deficiencies or gaps and of providing for the quick processing of the collected data." They agreed that the Air Force was already several years behind in certain aspects of intelligence and that "any delay in solving current problems caused by a study of the period ten years ahead would be regrettable."[15]

In the final report, that of the Social Sciences Panel, Dr. John W. Gardner, chairman, pointed out that whenever human beings were inserted into a system "you have to reckon with human capacities and skills." For example, landing flaps and wheel controls were once so similar and so placed that pilots, in landing, oftentimes reached for the controls and instead of retracting their flaps raised their wheels. Again, the mixture control in World War II aircraft moved forward in some cases and backward in others, often with disastrous results. Accidents caused by these and similar poor designs often were called pilot error, Dr. Gardner said. His panel believed they were more appropriately classified as engineering-design error. "It will never be possible to eliminate human errors," he said, "but human engineering research is now sufficiently advanced so that there is no excuse for equipment which invites human errors."[16]

---

*This panel was still called the Intelligence Systems Panel at this time.

See notes on page 185.

Following the oral presentation of the reports, the secretariat reproduced and distributed them "as a first step in the preparation of a new *Toward New Horizons*" and the Executive Committee discussed possibilities for proceeding with the work.[17] Since time was a vital factor, the committeemen agreed that panel chairmen should solicit the help of promising young scientists, either from their panels or from outside the board, to do the actual writing of the final report under the supervision of senior SAB members. This would reduce the work of the senior members and allow them to complete the smaller and newer version of *Toward New Horizons* within several months. Dr. Getting proposed that there "be some military people around to stimulate ideas in the operational areas," and this, too, was accepted. The meeting closed on the intent to keep the study group as compact and young as possible and set it to work full-time for about two months in the summer of 1954. The board would then edit their work, with panel chairmen assuming responsibility for the final form of each report.[18]

By June 1954, many of the executive committeemen had had second thoughts on the project and decided it was a wasteful venture on various counts. The idea of using young scientists sounded attractive but, as Dr. Millikan pointed out, the study was "not the kind of thing that can be turned over to youngsters . . . [and] to begin with you cannot find such youngsters." Dr. Hastings agreed, noting that it meant "recruitment, security, and trying to convey to them our post-graduate training." But the major objection, pointed up by Dr. Soderberg, was that the work duplicated one of the major purposes of the board—"new ideas." To single such a task out for special treatment "somehow suggests that the role of the board has never been fully understood."[19]

These opinions triggered a flood of objections. In Dr. Getting's opinion, "the feeling [was] universal that anybody . . . foolish enough to make a ten-year prognosis [at that time] should not be on the board." Dr. Sherwin suggested that RAND and other full-time contract non-profit organizations could "do jobs of this sort as well as the SAB." Dr. Millikan said that his panel "gave, in our best judgment, the fields that we felt would be fruitful in the next five years [and] I can see nothing that would be added by writing a monograph." Finally, Major Whitcraft reminded the Executive Committee that Dr. von Karman's "first reaction was very lukewarm," and

See notes on page 185.

that von Karman had gone along with the project in the first place simply "because he thought that was what the board wanted."[20]

As a consequence of this barrage of objections, the project was dropped.   At the September 1954 meeting of the Executive Committee, panel chairmen amended as necessary their earlier reports, and the secretariat gave them limited circulation within the Air Force, carefully observing Dr. Soderberg's injunction to "see that they are properly labeled and that the term *Toward New Horizons* [was] eliminated."[21]   The board wanted no one to mistake these cursory statements as an effort to re-do the study.   The committee also endorsed Dr. Doolittle's proposed explanation to General Putt (sent on September 29, 1954) why a new edition of *Toward New Horizons* would not be the best approach in meeting ARDC's requirements:[22]

> Research and development generally has been intensified and horizons . . . extended beyond the compass of an individual or a workably small group of individuals. At the same time, many new organizations have come into being, such as the Office of the Deputy Chief of Staff/ Development, the Office of the Assistant for Development Planning, and the Air Research and Development Command. Projects such as RAND and LINCOLN have been established, and other groups such as the Control Systems Laboratory have been formed to meet this need in somewhat more specialized areas.

> Moreover, there is at this time no mass of suddenly available scientific data as we obtained from Germany and Japan at the close of the war, or of suddenly available scientific talent as was provided by our exploding demobilization of 1946. In short, expanding horizons have greatly increased the scope required of such a work, while at the same time new groups, new laboratories, and other organizations have become responsible for all or part of the task you described. The conclusion seems inescapable that a re-do of *Toward New Horizons* would require a much greater effort with a much less useful result.

Recognizing the "long-term look" to be a continuing SAB role, Dr. Doolittle promised General Putt that the board would give the Air Force as much help as it could on the matter in the future.   Specifically, he had asked SAB members to report any significant development in their specialties "which portend opportunities for Air Force exploitation."   He would forward these immediately to the proper USAF agency.   Also, the Executive Committee would consider the possibility of calling a third meeting of the full board each year during which members

See notes on page 185.

could explore "the trend capabilities of those technical areas which will contribute most to the development of Air Force equipment in the next ten plus years."[23]

See notes on page 185.

CHAPTER SEVEN

EVE OF NEW CRISIS

The environment in which the board operates has changed appreciably since the end of World War II. The technical content of the Air Force responsibility has increased, leading to a USAF research and development program several times larger and many times more complex than it was in 1945; new agencies have been created . . . for the specific purpose of managing USAF R&D activities; interest in scientific matters has shown a continuing marked increase within the Air Force reflected not only in larger numbers of personnel directly or indirectly involved but also in an expanding level of educational, professional, and intellectual experience. As a result of these changes, the job of advising the USAF on scientific matters and the broadening personal contacts which this implies, has become an increasingly demanding and complicated task.

——Courtland D. Perkins*

In September 1954, Dr. von Karman notified the Chief of Staff, General Twining, that he could no longer, for his health's sake, continue both his AGARD and SAB chairmanships and asked to be relieved of the latter by the end of the year. "I feel that it was a great privilege to serve the Air Force," he wrote. "I cherished the memory of many episodes

*In memo to General Twining, April 3, 1957.

of my working for General Arnold. I was grateful for the support I received from his successors and I especially appreciated your confidence . . . and friendship." In accepting his resignation, General Twining offered him the post of SAB chairman emeritus in order "to maintain upon the SAB rolls . . . the name with which it had so long and so proudly been identified." Dr. von Karman accepted, continuing in that position until his death in 1963.[1]

Dr. Doolittle, speaking for the SAB Executive Committee, recommended to General Twining that Dr. Kelly take von Karman's place.[2] Dr. Kelly accepted, assuming office in January 1955.* "The job is tremendous in its possibilities . . . and I expect to work hard at [it] and make it my chief non-paid military job," he told the Executive Committee.[3] Unfortunately, after guiding the board through its 1955 major meetings, he found it necessary, because of the press of other affairs, to relinquish the chairmanship late that year and to retire completely from board membership at the end of the next. Again, board leadership devolved on Dr. Doolittle, except that this time he accepted General Twining's offer to assume the chairmanship officially. Doolittle informed Kelly that he thought to hold the position only until he could "return the baton . . . to you."[4] As it turned out, he remained chairman for the next three years.

General Twining concurred in the Executive Committee's proposal to offer the now vacant vice chairmanship to Dr. Stever, who accepted but had to postpone assuming office until August 1956 when he completed his year's tour as Air Force Chief Scientist. Meanwhile, General Putt succeeded General Craigie as Deputy Chief of Staff/Development and SAB military director, occupying that post for approximately the same period that Doolittle served as chairman.

In mid-1956, the Executive Committee acted to give officers in key research and development positions greater participation in their councils. The Chief of Staff approved the revision to the SAB regulation (AFR 20-30) to extend ex-officio membership in the board and Executive Committee to the Chief Scientist, the ARDC commander, and major directorate heads of the Deputy Chief of Staff/Development. The revision also enhanced the prestige of SAB membership by granting mem-

---

*Lt. Col. Floyd J. Sweet became board secretary at this time and Mr. Hasert accepted the new secretariat position of technical director.

See notes on page 185.

bers (and their invited guests and consultants) the same social and facility courtesies accorded lieutenant generals.*[5]

For the most part, the general board meetings during 1955-1957 proceeded along guidelines adopted in 1954, convening twice yearly (spring and fall) and at stations intimately involved in the subjects explored. One exception occurred in 1955 when Dr. Kelly, in keeping with the board's promise to General Putt to consider "the interaction of technology and warfare in the time period 1965 and beyond," held a third meeting on 13-15 June at a resort lodge in the Minnesota woods. Dubbed a "think session," the meeting was "generally considered successful" and board officers thought briefly of scheduling such third meetings in future years.[6] As the record of the meetings reveals, however, the idea was soon abandoned; from 1956 the full board met only twice yearly.† However, the meeting record also shows that the Executive Committee found certain features of the Minnesota session sufficiently valuable to incorporate them, wholly or in part, in the fall meetings of the ensuing three years. Here they sought to keep the formal agenda to a minimum and allow members a maximum of time to exchange ideas or, as one secretariat officer noted, to "provide a common meeting place in an abundance of trees, lakes, [and] good food . . . [where everyone was] encouraged to think out in the wild blue yonder."[7]

An early 1957 self-analysis served to point up various SAB management and procedural dissatisfactions expressed during 1955-1957. Prof. Perkins, then serving as the Air Force

---

*This practice had actually been followed for several years, but apparently had never been officially recorded on the Air Force precedence list. In October 1955, General Putt called to General Twining's attention the fact that the SAB secretariat had, in the past, "quoted an earlier reading to the effect that 'as consultants to the Chief of Staff, SAB members rank with, but after, active lieutenant generals,' but we find that this statement now has no basis in the current precedence list so that a new determination is necessary." Putt, at this time, asked that the assimilated rank continue to be accorded SAB members, noting that it had served well in such matters as "arranging appointments, expediting and facilitating visits to Air Force installations . . . and, perhaps most of all, in impressing our members with the importance we attach to their services." Twining approved continuance of the practice on October 27, 1954, and it was subsequently written into the SAB regulation and practiced throughout the following years.

†See Appendix H for the dates, location, and subject of SAB general board meetings from 1946 through 1964.

See notes on page 185.

Chief Scientist, suggested the study to General Twining at the behest of the Executive Committee.[8] Twining approved, inviting Perkins to head up a committee that included SAB members Maj. Gen. James McCormack, USAF (Ret.), and Dr. Valley and consultants Mr. Peter J. Schenk and Mr. Walkowicz. They submitted their report to General Twining on April 3, 1957, and, soon after, discussed it with him at a luncheon meeting.

In a prefatory statement, the Perkins Committee explained its views on the impact which the changing times had had on board activity.* They also noted the changes wrought by the years on board members, pointing out that "the 'eager beavers' among the scientific community whose ages ranged from 25 to 35 and who participated so actively in the scientific developments of World War II have not been replaced by a 'new young group' equally knowledgeable about problems of defense."† Also, many board members had "shifted in their day-to-day activities from creative research work to more administrative and managerial functions." All of these changes, the committee believed, emphasized the need "to strike a balance in board composition between the capability for advising the Air Force on purely scientific questions versus those involving broader policy matters of research and development programs, organization, and management."[9]

The Perkins committee strove essentially to uncover deficiences that continued to depress board talents. The group observed that a successful board depended ultimately "on its being composed of good people, properly employed." Consequently, the board had to improve on its ability to "impart to the members a sense of satisfaction in making significant contributions to the Air Force." To effect this, the group saw a need for (1) improving communication channels between SAB members and Air Force officers on the working-level, (2) bringing more senior Air Force officers into closer contact with board activities, (3) maintaining a better membership balance between "creative young scientists [and those] of broad management-oriented background," and (4) enhancing the prestige of secretariat duty.[10]

Because it was overtaken by events, the Perkins Report had little immediate impact on board organization and prac-

---

*See quotation at beginning of this chapter.

†A 1958 secretariat study revealed that the average age rose from 43.9 years to about 48 in the 12 years since the board's founding.

See notes on page 186.

tices. However, SAB officers in future years implemented, or attempted to act on, most of its recommendations in one way or another. For example, the committee proposed only one change in basic SAB organization. It felt that the board should have a "Policy Panel," staffed with eight to ten senior members to handle major policy matters, as distinct from normal administrative affairs which the Executive Committee could continue to handle. Dr. Doolittle, while agreeing with the concept, deferred action on it, choosing instead to rely on the recently expanded Executive Committee to continue to handle all policy and administrative matters. Ultimately, however, under different names and composition, groups formed within the basic SAB structure that, in effect, achieved the ends the Perkins Committee sought in its policy panel proposal.*

Except for one adjustment, panels—in size, number, and areas of interest—remained essentially unchanged through 1955-1957. The one significant revision concerned the Nuclear Weapons Panel. As noted later in this chapter, the panel, under Dr. von Neumann's chairmanship, performed several extremely valuable studies from the time of its establishment in 1953 through the summer of 1955 and then lapsed into comparative inactivity because of von Neumann's illness. In April 1956, after prevailing upon Dr. Teller to chair it, board officers reconstituted the panel. Calling it simply the Nuclear Panel, they assigned it responsibility for all nuclear matters—nuclear weapons, reactors, propulsion etc. As finally manned in September 1956, the membership included former Nuclear Weapons Panel members Bradbury, Griggs, Kistiakowsky, Scoville, and Dr. Herbert F. York, and former Propulsion Panel members Dr. Alvin M. Weinberg and Mr. Gale Young.† Dr. Mark M.

---

*The board implemented immediately one Perkins Report proposal: The 1956 revision of the SAB regulation, the Report noted, "eliminated . . . the right of free communication between the SAB secretariat and all echelons of the Air Force." Formerly, the regulation read that "Air Force activities desiring the consultative services of individual board members will forward requests to the Secretariat." The change, in an effort to make the process more orderly, directed that they be sent "through normal channels" to the secretariat. The Perkins Committee found that field officers had misconstrued the change to mean that SAB did not want to be bothered with their small affairs. On April 17, 1957, the board, in a one-paragraph revision of the regulation, authorized field officers to once again submit directly to the secretariat.

†Dr. von Neumann also accepted membership, but ill health forced him to resign from the board at the end of 1956.

Mills, chairman of the Fuels and Propulsion Panel agreed to serve as liaison member "to insure a close tie-in" between his and the Nuclear Panel.[11]

The most succinct resume of panel assignments and accomplishments during the middle-1950's came from the panel chairmen themselves in the spring of 1957. Submitted on request of Dr. Doolittle, the papers sought to inform members on how the separate parts of the board had fitted into the whole over these years and to explain panel philosophies and aspirations.[12]

Dr. Clayton S. White reviewed the work his Aeromedical Research Panel had done on the problem of acquiring qualified personnel for aeromedical research and (in conjunction with the Nuclear Panel and aircraft nuclear propulsion study groups) on the biologic problems associated with the employment of nuclear propulsion and nuclear weapons. Concerning the future, Dr. White felt certain that his panel, as well as the SAB, could always find plenty to do if it were creative and aggressive enough, noting that

> information, both basic and applied is—from the point of view of an advisory group—as important as is 'know how' in industry. I believe in the effectiveness of the simple concept that if a lad keeps busy every hour of the working day, he can well leave the upshot of his education to itself; e.g., if the Aeromedical Panel through its several meetings and discussions can demonstrate and continually develop a knowledgeability and competence, the range of its activities will not be limited by the existence of problem areas in which it is requested to work . . . [Future] Panel activities . . . for sure will fall into at least two categories; i.e., (1) aiding and participating in the exchange of available information between disciplines whether biologically or physically oriented, and (2) clarifying and categorizing the need for unavailable information and developing the conceptual attack into unexplored fields whether these be basic, applied, or both.

Specifically, Dr. White noted that the Aeromedical Research Panel intended to continue work on such areas as toxicology of new fuels and propulsives, escape and survival problems posed by new high speed aircraft, and the biological aspects of atmospheric flight. Also, the panel intended to continue to press for the creation of adequate Air Force radiobiological research facilities. Finally, it intended to continue its interest in protective construction at air stations and in the medical handling of mass casualties.[13]

Dr. C. B. Millikan cited the contributions his Aircraft Panel made to the Air Force intercontinental ballistic missile pro-

See notes on page 186.

gram, noting that "after some year's delay the recommendations have been put into effect practically in their entirety." He also noted the work the panel had done in support of the B-58 program, the aircraft nuclear propulsion program, the project for developing a 100,000-pound payload logistics carrier and tanker, the AVRO circular wing aircraft, and the X-15 research vehicle. On the panel's future plans, Dr. Millikan expressed his personal philosophy "that a proper and useful function of the Panel is to suggest general areas of long-range future research, but that its greatest effectiveness will probably continue to be demonstrated when it is asked to study and make recommendations concerning specific problems which are currently under discussion and coming up for decision by the Air Force."[14]

Dr. Getting's paper surveyed the history of the Electronics and Communications Panel since 1952. On electronic countermeasures, which had been the panel's major area of interest since the early 1950's, he noted that the science had developed so rapidly that the enemy could now quite easily jam essential communications. The panel's contributions toward minimizing this threat had been to acquaint the Air Staff, ARDC, and the operational commands with the technology involved and to make recommendations "which led directly to the establishment of a Quick Reaction Capability (QRC), methods of handling quick reaction operations requests, and establishment of QRC . . . officers in all operational and supply commands." Self-jamming by friendly equipment had also been a subject of panel concern for many years. For a long while, panel recommendations for studying the problem jointly with the other services "were frustrated by considerations other than technical." Fortunately, a RAND study, combined with actions initiated by panel members who also served as members of the Army Chief Signal Officer's Research and Development Advisory Council, led to the establishment of Project Monmouth which, in turn, "brought about a broader recognition of this problem in both the Army and the Air Force." As a result of this effort, the services had achieved a more orderly assignment of frequencies.[15]

The Electronics and Communications Panel also contributed to the establishment of a reliable European aircraft control and warning system. The lack of such a system (and poor communications generally) within the NATO complex greatly

See notes on page 186.

minimized the U.S. Air Force's materiel and personnel investment in those forces. In conjunction with the Standing Group
of NATO and the Air Staff, panel members made a thorough
study of NATO needs in this regard. General Craigie, while
Deputy Chief of Staff/Development, and Dr. Chalmers W. Sherwin, while Chief Scientist, helped with a similar study. As a
result, in March 1954, the panel recommended the establishment
of an electronic technical laboratory under NATO, staffed by
European scientists, to develop an intregrated control and warning system with indigenous equipment. By 1957, the laboratory was in operation and had made "significant progress."[16]

Finally, Dr. Getting noted his panel's contributions to improving or amassing data on such projects as infrared applications, tactical air integrated data handling systems, naviga-
tional aids, radar reconnaissance, and ballistic missile defense.
Future goals of the panel included "continued investigation of
the major problems still outstanding in [electronic counter-
measures], communications security, weapon systems electronics, and frequency planning."[17]

Dr. Draper also injected a historical summary into his
Explosives and Armament Panel presentation, noting that
these were "fields of primary importance for the SAB during
its first meetings in the mid-40's and have continued to hold a
position of this kind throughout the history of the board." In
the early days, attention had concentrated mostly on guns and
on bombs with chemical warheads. Later, "rockets and guided
missiles started to take the center of the stage and have held it
ever since with increased performance, size, and the much
greater destructiveness that goes with nuclear warheads."
Meanwhile, fire control underwent an evolution that roughly
paralleled the changes in the power of the weapons. In the
early years, aircraft fire control employed visual tracking for
rocket and gun weapons and operated at ranges of not more
than a few hundred yards. Now, however, the existing and
contemplated systems applied radar or infrared tracking,
achieved maximum ranges of several miles, and were designed
to operate with greatly improved guns, rockets, and guided
missiles. As for bomber defense, it changed "from a matter of
gun turrets adapted principally for limited rearward coverage
to elaborate electronic systems providing countermeasures, eva-
sive maneuvers, and all angle protection with guided missiles."

See notes on page 186.

Radar bombing systems had now largely replaced optical systems.[18]

Throughout these rapidly changing times, Draper said, his panel had "advocated flexibility and balance in Air Force weapon systems," basing this position on the assumption "that our potential enemy is determined and resourceful and will certainly be well informed on our combat equipment, so that we must be prepared to counter any one of many types of attack and also be ready to retaliate with an effective element of versatility to supplement the power of our weapons." In short, the panel believed "that we should be prepared to fight not only all-out nuclear, bacteriological-, and chemical-agent wars but also 'brush-fire' actions in which guns and nonatomic explosive munitions will be the best tools."[19]

Dr. Draper listed a number of the major recommendations that the panel had made over the years. It had

repeatedly called attention to the performance benefits that are possible in solid propellants for rockets . . . improvements [that] are just now being realized in research projects. In the field of air-to-air rockets and guided missiles, the desirability of warheads large enough to give lethal results without the necessity for direct hits has been stressed. Attention has also been repeatedly called to the need for fire control equipment with maximum freedom from initial positioning limitations in order that dependence on ground control may be minimized during combat. The very great crowding of radio and radar channels and possible difficulties from electronic countermeasures have been a continuous source of concern [leading to the panel's support of the development] of self-contained systems for navigation, control, and bombing. In order to assist in the achievement of such equipment, the Panel has been very active in encouraging the development of inertial systems to meet as many combat requirements as possible. Similarly, passive tracking systems using infrared or other electromagnetic radiation have been watched closely and given assistance when possible. The Panel has recognized the effectiveness of ballistic missiles for medium-range and intercontinental bombardment and . . . assisted in making the decision that . . . resulted in the Atlas and Thor projects. Panel members served on various [SAB special committees] whose reports . . . led to the MB-1 air-to-air nuclear-warhead rocket and the initiation of guided missile developments to follow this rocket as Air Force weapons. [Finally, the] great increases in air-to-air combat ranges that depend on the larger lethal radii of nuclear warheads . . . led members to suggest revolutionary changes in aerial combat equipment having preset inertial guidance for high-maneuverability and high-Mach-number guided missiles as the damage-inflicting agents.

In his opinion, nothing on the scientific horizon hinted that armament problems would "decrease either in number or com-

See notes on page 186.

plexity" in the future, and if the Air Force were to preserve "a good technological margin of superiority" over its possible enemies it had to continue to maintain "a strong concentration of attention and effort in this field."[20]

Dr. Mills reported that his Fuels and Propulsion Panel had encouraged development of supersonic and subsonic compressors "in order to achieve supersonic turbojet propulsion," maintained a continuous surveillance of super fuel developments, and kept in close touch with novel power plant ideas. "The interplay of the new fuels, and the new engine cycles, leads to a very complex picture of propulsion possibilities," he noted. The panel also encouraged greater emphasis on the science and art of materials development, recommending "support of basic and applied materials work across a broad front."[21]

Dr. Kaplan, Geophysics Research Panel chairman, applauded recent Air Force fund increases to geophysics research and development. However, he felt the overall allocation was still too low. Whereas most USAF research and development programs received industrial venture capital as well as Air Force funds, this was not true for geophysics. On most other counts, Dr. Kaplan was pleased with the progress of his panel's work and with the Air Force's reception of it, noting that "at the working level . . . the recommendations and suggestions of the panel have always been accepted in a cooperative spirit, have been given serious attention by senior personnel, and have been incorporated in the research program and carried forward successfully." Specifically, on recommendation of his panel: ARDC had assumed total responsibility for the whole field of meteorological instrumentation; the Cambridge Research laboratories had agreed to consult the Geophysics Research Directorate before making any major programming decision in the geophysical field, had initiated studies on an automatic atmospheric reporting and forecasting system and on polar meteorology, and had continued the Ice Island project; and the Air Force had begun work on a space satellite to serve as a platform for geophysical and other types of observation.[22]

The Nuclear Weapons Panel had barely formed in 1953 when General Thomas D. White, acting Vice Chief of Staff, had asked it for an estimate on the size and type of nuclear weapons which the Air Force might employ over the next six to eight years. Design of future delivery vehicles and attack strategy and tactics depended in great part on the characteris-

See notes on page 186.

tics of these weapons, yet the Air Force had no reliable information on them.[23] Replying in mid-1953, the panel set forth the bomb yields "that were likely to be achieved in the next five years or so in the major weight classes."[24] The report had a vitally important impact, both in advancing the Air Force development program and in triggering approval of the ballistic missile program. As General Putt noted, this particular report must always be regarded as "one of the most outstanding and important examples of board influence [and] effectiveness."[25]

The next year, the panel threw even clearer light on the possible range of applicability of thermonuclear weapons, predicting that "yields as high as 1 megaton per ton of weight were possible."[26] In Dr. von Neumann's words, these findings confirmed "that a thermonuclear weapon could be incorporated in a ballistic missile" and led to the Presidential decision which resulted in the shift to the dynamic ICBM program, headed by General Bernard A. Schriever, which produced America's operational missile force.[27]

After the panel was reconstituted as the Nuclear Panel in 1956, it investigated the fall-out hazards which might arise from nuclear propulsion and accidental nuclear warhead detonations and concluded that neither presented insuperable problems. It also supported the philosophy that both theoretical work and actual tests were necessary if the Air Force were to secure adequate technical information on high-altitude nuclear burst effects.[28]

Dr. Macdonald reported that his Reconnaissance Panel had concentrated on pre-hostilities reconnaissance requirements in the past and planned to continue emphasis on this area. At the same time, members intended to explore "the existing potential and projected research and development programs in the combat reconnaissance field." Concerning the value of his panel's efforts, Dr. Macdonald pointed out that "as, perhaps, with most panels, the scope of effectiveness of [our] activity must take into account an informal role achieved through direct interpersonal relations with USAF officers."[29]

Dr. Dael Wolfle, Social Sciences Panel chairman, noted that this applied equally to his panel, that "individually and collectively, members . . . have served as advisors . . . , no record has been kept of the subject matter or results of these con-

See notes on page 186.

sultations, [and] whatever merit they have had has been reflected in the work of the human resources research activities
of the Air Force." Since this method of operation had enabled
them to pass many specific suggestions on directly, most of the
panel's formal reports had dealt with broad organizational policy recommendations intended to improve Air Force human
factors engineering programs. Essentially, the panel directed
its efforts toward establishing valid procedures for evaluating
the outcome of these programs and for introducing consideration of human factors aspects much earlier in the design of
future weapon systems. That all of this was "necessarily a
slow business" should not be surprising, he concluded. His
panel dealt with such imprecise subjects as personnel selection,
training, morale, social organization, and human relations—
"all topics that are the everyday operating responsibilities of
the Air Force or any other organization, and all topics on
which individual experience . . . may carry greater weight than
do research findings."[30]

See notes on page 186.

PART III

CHALLENGE FROM BEYOND THE HORIZON

. . . in 1945 we suddenly knew we had arrived at a new age—the Atomic Age. By the early fifties, it was clear that the Missile Age was upon us—and before that decade was out, we had reached the Space Age. The most startling things about this bursting forth of technology is the increased pace of change. Technology seems to be rushing headlong into the future. The newest succeeds the new. It's a common joke among engineers that if you know how to build a thing, it's obsolete. In fact, if you don't stay alert today, you may miss an age as it goes past.

——*James Ferguson*, in speech at Purdue University Convocation, April 13, 1964.

CHAPTER EIGHT

WITH SURVIVAL THE GOAL

> . . . if we talk about the present
> then I know that all the experts
> are in the Air Force and they
> know what they are doing and
> they don't need a glass. When we
> talk about the future then there
> are no experts anywhere and there-
> fore, also, people like ourselves
> qualify.
>
> ——Edward Teller*

Sputnik I, the 184-pound satellite which the Russians shot into orbit on October 4, 1957, confirmed the scientific and military communities' oft-repeated warning that the United States lagged in space research and development. While a few high government officials unwisely chose to publicly deprecate the launch as "a neat scientific trick," or a sort of "outerspace basketball game," worried DOD agency heads promptly acted to assess the Russian achievement in terms of its threat to U.S. security.[1]

Within the SAB, one officer felt the new crisis sufficiently grave to suggest that the board consider the coming months as ones "of national emergency, when survival may be determined."[2] And, as in previous crises, the Air Force quickly solicited the SAB's counsel. After attending the series of high-level conferences held between October 8-15, Secretary of the Air Force James H. Douglas called for a study of steps the Air Force might take to assist in countering world reactions to the Soviet space accomplishment. Lt. Gen. Samuel E. Anderson, ARDC commander, assumed responsibility for the study and, with Dr. Doolittle's help prevailed upon Dr. Teller of the SAB Nuclear Panel to spearhead it.[3] Dr. Teller immediately

---

*During a discussion at a December 1957 SAB meeting.

See notes on page 186.

assembled an ad hoc committee of SAB members, industry experts, and ARDC technical advisors who met on October 21-22, 1957, and submitted a report through channels to Mr. Douglas on the 28th. They stated their conclusions and recommendations briefly and to the point. The United States had slipped behind the Soviet Union in the technological race because of complacency and swollen bureaucracy. As a result, neither civilian nor military research and development agencies had been able to establish stable, imaginative programs. To correct the situation, the Teller Committee urged that the organization and management of ballistic missile and space flight programs be consolidated and simplified from the Office of the Secretary of Defense down through the services. The government then had to give these programs top priority "without reservation as to time, dollars, or people used."[4] As for the manner in which the nation's leaders ought to react to the Soviet success, the committee felt America should honestly admit it had been eclipsed and honestly recognize the reasons for it. The members warned that "an attempt to counter the sobering effect of Sputnik . . . by a spectacular, but technically superficial demonstration would be to seriously and perhaps fatally deceive ourselves as to the gravity of the present technical position of our country."[5]

The SAB underwrote expenses of its members who served on the Teller Committee and would have proudly accepted credit for its work. But the Teller Report was never considered a SAB product in the official sense. However, by a fortuitous circumstance, SAB had formed a space study group of its own several months before Sputnik. In November 1956, the Fuels and Propulsion Panel had recommended that the problem of national defense in cislunar space be studied on as broad a basis as possible.[6] General Putt, finding the idea appealing, requested the SAB to form an ad hoc committee on advanced weapons technology and environment. This was done in May 1957 with Dr. Stever as chairman and Drs. Kaplan, C. B. Millikan, Mills, William H. Radford, Simon Ramo, and White members. Studies under way at Ramo-Wooldridge, RAND, Western Development Division, and Headquarters USAF gave promise that the ballistic missile program would soon produce vehicles capable of operating in space. Since this would have a "severe impact" on military operations, Putt considered it imperative that the Air Force "keep abreast of the latest thinking in these areas and . . . be immediately informed of potential break-throughs." Consequently, he asked the committee to assess current technological knowledge on the subject then advise the Air Force of the direction it should take to explore the new environment and to study the weapon systems required for operating there.[7]

The committee began its meeting in late July 1957 and submitted a report to Dr. Doolittle just before the first Sputnik shot. Doolittle completed final coordination and amendment and forwarded it to the Chief of Staff, General White, on October 9, five days after Sputnik.[8] The report recommended that the Air Force strongly support pure research on matters of space exploration. This would have two major benefits. It would provide the Air Force with new information applicable to space flight and, at the same time, ready the Air Force to assume what would appear to be its logical future mission of performing space logistics "analogous to the Navy's logistics capability in bringing scientific data back from the Antarctic." The committee believed the Air Force should act promptly on its recommendations, "the urgency here [being] substantially above that of the average problem submitted to the SAB."[9] These words expressed the view of the SAB as a whole and served as SAB's first official declaration of concern over the lead the Russians held in space technology.

Members of the Stever and Teller Committees present at the first full board meeting after the Sputnik crisis (held December 4-6, 1957, at Chandler, Arizona) reviewed their reports in concert and discussed further thoughts they had had on the subject in the intervening weeks. They agreed that "Sputnik and the Russian ICBM capability [had] created a national emergency." They also agreed on the course the Air Force should pursue to meet the emergency. Accordingly, Dr. Teller and Prof. Griggs joined with the Stever Committee (who were all in attendance except Dr. Ramo) to form a new ad hoc committee on space technology chaired by Dr. Stever.[10] They then issued a new statement which Dr. Doolittle submitted to General White a few days after the meeting, noting that it reiterated "previous statements made by SAB members which we feel need reemphasis in light of the critical post-sputnik situation now existing."[11] The terse report advocated prompt and vigorous action on six fronts:

See notes on page 187.

(1) Obtain a massive first-generation IRBM and ICBM capability as soon as possible. (2) Establish a vigorous program to develop second generation IRBM's and ICBM's having certain and fast reaction to Russian attack. (3) Accelerate the development of reconnaissance satellites. (4) Establish a vigorous space program with an immediate goal of landings on the moon. (5) Obtain as soon as possible an ICBM early warning system. (6) Pursue an active research program on anti-ICBM problems. The critical elements are decoy discrimination and radar tracking. When these problems are solved, a strong anti-ICBM missile system should be started.

Soon after these recommendations went forward, General Putt called on the SAB to review continental air defense programs. The new show of Russian technological progress "reemphasizes the need for a thorough examination of our defensive requirements to meet these threats in the next 10-year period," he said.[12] He listed the "spectrum of possible threats," as he saw it, and in January 1958, a SAB ad hoc committee under Dr. Sherwin set out "to determine the minimum number of integrated systems necessary to meet all threats." General White was disappointed in the committee's first findings, sent him in draft form in mid-1958. As one secretariat officer expressed it, the Air Force had hoped that some "new and radical air defense techniques" would emerge from the committee's labors.[13]

The committee expanded its search after members Sherwin, Ernst H. Plesset, Radford, Richard C. Raymond, and Valley met with General White in late August to gain a better idea of his views.[14] Their final report, completed in January 1959, noted various "serious gaps" in air defense programs and stressed "the need for accelerated and specialized research and development" in such areas as ICBM warning and anti-ICBM missiles.[15] After the committee disbanded, General Putt stated that he felt the members had accomplished their primary purpose and congratulated them on their work.[16] There seemed to be a lingering note of disappointment among all concerned, however. Unlike the Valley Committee's air defense studies following the first Russian atomic explosion in 1949, the Sherwin study was unable to stimulate any "new approach" to the problem. Subsequent experience made clear why this was an inevitable result. While the Air Force could and did accelerate the development and installation of missile warning systems, the state of the art was not sufficiently advanced to enable anyone to conceive reliable systems for intercepting and destroying missiles.

See notes on page 187.

In short, the problem of "aerospace" defense had become vastly more complicated than was envisioned prior to the arrival of ICBM's on the scene.

Concurrent with its other labors immediately following Sputnik, the SAB undertook an assessment of ARDC's ability to handle the programs whose need for rapid completion could now be equated in terms of national survival. The Ridenour Committee of 1949 had suggested that SAB take a new look at Air Force research and development once the recommended organizational changes were implemented and in operation for a reasonable period. General Thomas S. Power, ARDC Commander in the mid-1950's, spoke of the advisability of such a study from time to time but never officially requested it. His successor, General Anderson, had been Director of the Office of the Secretary of Defense's Weapon System Evaluation Group and had attended SAB meetings and fully understood the board's purpose and operating methods. He informed Dr. Doolittle and General Putt privately soon after he took command of ARDC that he favored a SAB re-study and would ask for it as soon as he became familiar with the many facets of his organization. Later, after he had found that many of ARDC's programs were frozen for lack of funds as a result of the Eisenhower Administration's decision to hold the national debt ceiling under $275 billion, he submitted a formal request for the study. The SAB Executive Committee met with General White on the matter in September 1957, and obtained his verbal approval to proceed. In accordance with General Anderson's wishes, the Executive Committee enlisted as many members of the Ridenour Committee as could lend their services to the new group and Dr. Stever agreed to chair it.

On November 21, 1957, the new ad hoc committee on research and development, comprised of about 20 members, met in the Pentagon for the first time. The same day, General White formalized its mission, now made more urgent by the Sputnik crisis, asking for "an impartial and searching review of the organization, functions, policies, and procedures of the Air Force and ARDC in relation to the accomplishments in research and development over the past seven years" and for "recommendations as to how we can do the job better in the future." Soon after, General Anderson informed all ARDC units that the survey was no ordinary one and enjoined them to be completely frank in answering all questions. They in turn established special project officers to insure compliance.[17]

The committee met for several days in December 1957 with agencies in the Washington-Baltimore area, spent nearly two weeks in January visiting agencies in the west and mid-west, and wound up their purely investigative labors with a similar tour of eastern agencies in mid-February. After spending some four months drafting a report, the committee submitted it to Dr. Doolittle on June 20, 1958.[18]

The committee acknowledged the gains that the Air Force had made as a result of the 1950-1951 reorganization. For example, the creation of ARDC and the Office of the Deputy Chief of Staff/Development had brought research and development officers into the highest policy-making and planning councils of the USAF for the first time. As a result, the nation's scientific resources, particularly those of the universities, were brought to bear more effectively on critical USAF problems. Also, research and development personnel, initially in critically short supply and widely dispersed, had been gradually assembled under the new organizations. Though these gains had been stimulated by the Korean War, the alarm over the growth of Soviet military power, and the mounting evidence of Soviet scientific and engineering prowess, the committee believed that "the timely reorganization of the Air Force research and development activity did permit the Air Force to make an impressive gain" from the additional monetary support obtained in the early 1950's.[19]

Unfortunately, the Stever Committee pointed out, these gains were "partially vitiated [by] limited budgets and excessive administrative controls . . . compounded by some evident reservations within the Air Force about either the capability of the research and development organization or [its] importance." Toward the end of the Korean War, just when improvements in research and development were beginning to bear fruit, the government acted to reduce military costs by relying more heavily on modern weapons and technology. While this should have brought additional increases in research and development funds, an opposite policy was adopted. As a result, the Air Force research and development budget leveled off, then declined. Also, stricter controls were adopted, with project decision often carried to much higher administrative levels than

---

See notes on page 187.

before. Finally, decision-making became more ponderous as did procedures for providing the resources—people, dollars, and facilities—required to get the research and development job done.[20]

To place Air Force research and development back on firm footing, the committee recommended a twofold approach. The Air Force should press harder to persuade higher authorities to lift the restrictions on resources on the assumption that Air Force research and development problems were not "inseparable from those . . . within the Department of Defense and the entire structure of government." Secondly, the Air Force should make specific organizational and management changes that were within its province of authority. These were intended to correct such weaknesses as duplication and over-control, restrictive procurement practices, faulty organization, and inadequate support of basic research.[21]

Air Force agencies greeted this Stever sequel to the Ridenour Report with mixed feelings. Lt. Gen. Roscoe C. Wilson, deputy to General Putt, wrote General Anderson on June 24 that the Air Staff was "in general agreement with the philosophies and principles set forth in the committee report," and thought that "many of the recommendations contained therein can and should be implemented at an early date."[22] ARDC, in a lengthy commentary on the report submitted to Headquarters USAF in early August, also agreed that many of the recommendations could and ought to be adopted immediately. However, it did not think that duplication, over-control, or inadequacy of support for basic research were as grievous problems as the Stever Report indicated. Consequently, ARDC did not concur in any extensive reorganization. Soon after, the Air Staff completed its own evaluation of the Stever Report and agreed generally with the ARDC review. It noted that the report was a most valuable stimulant and safeguard "against falling into archaic ways" but felt this was not an opportune time to apply sweeping change.[23]

No one, including the officers who wrote the ARDC and Air Staff critiques, seemed completely satisfied with these preliminary statements. As one result, General Anderson appointed a committee from his own staff in the fall of 1958 to reexplore certain aspects of the subject. In early 1959, this group expressed agreement with a number of important Stever Committee recommendations.[24] Later in that year, General Schriever suc-

See notes on page 187.

ceeded Anderson as ARDC commander and promptly initiated
further broad study. From this time through the April 1961
reorganization of ARDC into the Air Force Systems Command
(AFSC) and the Office of Aerospace Research (OAR), the Air
Force gradually applied many of the organizational principles
set forth in the Stever Report.

At the same time it formed special committees to handle
the diverse, immediately pressing problems which came its way
following Sputnik, the board enjoined its regular panels to step
up the tempo of their studies. The panels had always done excellent work whenever they "were stimulated by specific request," Dr. Doolittle told his panel chairmen, but in the face
of the Sputnik crisis he hoped the panels would "try to achieve
more outstanding jobs as a result of self-stimulation."[25] Meeting on the subject in December 1957, the panel chairmen agreed
that their groups had to meet more regularly than in the past
if they were to carry out Doolittle's wishes. As one member summarized the advantages here, "the more active the panel is, the
more closely we are in tune, and the greater the number of
problems which come to light, either spontaneously or through
requests."[26] They also agreed on the need for more cross-
fertilization among panels. Accordingly, the Executive Committee set time aside in future general board meetings for chairmen to brief the whole membership on their panels' current
activities.*[27]

As part of this effort to revitalize panel operations, all but
two of the nine panels underwent name changes in late 1957
and 1958 to better reflect their primary areas of interest. The
Aeromedical Research Panel became the Aeromedical/Bio-
sciences Panel. The Aircraft Panel changed to Aerospace Vehicles, indicating its interest in all manned vehicles, astronauti-

---

*Panel chairmen delivered such reports to the 1958 and spring 1959
board meetings. In June 1959, the Executive Committee tightened board
policy for handling controversial SAB reports, including restricting such
presentations "except in certain selected cases." Later, however, the
practice was again introduced and, still later, was again questioned, this
time for being too time-consuming. However, many members still felt
that the original reason for adopting the practice remained valid—that
the reports kept panels apprised of one another's activity and problems
and also "stimulated discussion and a desired degree of controversy."
The Executive Committee, in April 1963, adopted the compromise solution
by asking two or three panels to report at future meetings, the selection
depending on the general subject of the meeting and on the degree of
importance or interest of panel activity.

See notes on page 187.

cal as well as aeronautical.* The Fuels and Propulsion Panel and Communications and Electronics Panel dropped the first half of their names. Geophysical Research became simply Geophysics. Explosives and Armament changed to Guidance and Control. Social Sciences expanded its name to Psychology and Social Sciences.† Only the Nuclear and Reconnaissance Panels retained their old names.

---

*This panel changed first to Aero and Space Vehicles, but changed to Aerospace to conform with Air Force usage.

†In explanation of this name change, Dr. Fitts said the panel had experienced difficulty with the old one because "over the past several years the board has considered many problems to which the Social Science Panel could have made a contribution but no one thought of asking us." Panel members thought the new name would do a better job of indicating the areas in which the panel was competent. Too, the panel was usually comprised of psychologists as well as sociologists, economists, and political scientists. And psychologists did not regard themselves as social scientists.

CHAPTER NINE

A VERY EXPENSIVE PLACE

> We have had it impressed on us that space is a very expensive place; consequently, anything [the SAB] can do to channel our efforts toward the more fruitful objectives, to highlight potential scientific problem areas, or to indicate ways of getting past technical obstacles would be very much appreciated.
>
> ——Curtis E. LeMay*

The first significant physical change in the board after Sputnik came in the fall of 1958 when it underwent an almost complete change in management. Having served on the firing line of SAB affairs for nearly nine years, Dr. Doolittle deemed it advisable to pass his chairmanship and membership into new hands in November 1958.[1] General Putt who retired from the Air Force earlier that year accepted General White's invitation to succeed Doolittle. Meanwhile, General Wilson had replaced Putt as Deputy Chief of Staff/Development and SAB military director. Dr. Stever remained vice chairman.†

The new officers quickly espoused a move already introduced to abandon the policy of restricting membership to 50. They felt that for the board to adequately respond to the flood of projects now arriving it had to expand both in numbers and disciplines. Membership had declined to 51 by 1958—the lowest since the early days of the Korean War. Following the decision to expand, membership rose rapidly—to 67 in 1959 and increasing each year afterward to a peak of 88 in 1962.†† At

---

*In letter to the SAB chairman, June 22, 1961.

†Colonel Clyde D. Gasser succeeded Colonel George H. Duncan as board secretary at this time. The latter had succeeded Colonel Sweet the previous year.

††See Appendix A for annual membership totals, 1946-1964.

See notes on page 187.

the same time, the SAB used larger numbers of consultants; from only 36 in 1958, the yearly average increased to 98 in 1959, 69 in 1960, 72 in 1961, and 84 in 1962.*

The need for expanding board membership was first aired at an April 1958 Executive Committee meeting after Dr. Getting pointed out that there were not enough electronic experts on the board to carry out study requests in this area.[2] Discussion on the subject quickly broadened into a reconsideration of the overall board structure. To date, of course, the board had always been structured vertically, by technological specialties. It was now suggested (as it had been on several occasions in the past) that perhaps the board would operate more effectively if (1) it were reorganized along functional or horizontal lines or (2) permanent "cross panels" were created to complement and harmonize the work of the panels. At the close of the meeting, Prof. Perkins agreed to head a group composed of himself, Dr. Getting, Dr. Stever, and Dr. Valley to explore possible alternatives to the current organization.[3]

The Perkins group met in August 1958 at Woods Hole, Mass., and submitted a report the following month. Its major conclusion was that the complexity of modern weapon systems had created a situation where the SAB no longer had the breadth of competence to perform the studies requested by the Air Force. Consequently, experts from unrepresented technological areas should be invited to membership to bolster current panels and staff new ones. The group discouraged the idea of permanent cross panels, proposing that the board continue its current practice of creating special committees as necessary. Otherwise board members might become confused as to their prime areas of responsibility and cross panels mushroom to a point where they dominated the whole board.

To achieve greater rapport between the disciplines, the Perkins group suggested (as an alternative to "cross panels") that related panels be joined into "divisions." Specifically, four panels—Aerospace Vehicles, Guidance and Control, Nuclear, and Propulsion, with a combined membership of 21—would be formed into an Aero and Space Weapon System Division comprised of 20 to 32 members. In the electronic area, the

---

*To perform the great increase in administrative services required by this expansion, the secretariat grew from four officer and three clerical positions in 1959 to five officers and a clerical staff of six in 1962. See Appendix B for a roster of secretariat officers, 1946-1964.

See notes on pages 187 and 188.

current panel of six would evolve into a Communications and Information Handling Division of 29 to 45 members.* The group did not feel qualified to suggest changes in the remaining four panels—Aeromedical/Biosciences, Geophysics, Psychology and Social Sciences, and Reconnaissance—but observed that the chairmen of these panels might find the division concept attractive.[4]

The Executive Committee studied the report at the October 1958 board meeting at Puerto Rico, discussing it in the light of how the board might better organize to serve the Air Force and, at the same time, derive the fullest use of all members. The latter point appeared after one SAB officer reported that "recently enthusiasm [had] appeared to wane" in certain quarters of board activity because members felt that they were not being fully utilized.[5] The ensuing debate covered many views, "ranging from modification or reorganization of the present panel structure [along lines suggested by the Perkins group] to maintenance of the status quo." But the only actions which resulted were an authorization for an expanded Electronics Panel and an increase in the number of Executive Committee meetings from two to four each year. The latter move, Dr. Stever noted, would permit the committee "to maintain dynamic control of SAB operations much better than we have in the past and detect and deal with problems that arise more quickly."[6]

Carrying through on the October 1958 discussions, the SAB Chairman, General Putt, suggested that it might be better after all, "from a functional viewpoint," to reorganize the SAB "to embrace organization elements topically oriented to scientific disciplines such as Energy Transfer (in lieu of Propulsion), Environment Physics (in lieu of Geophysics or Astro Physics), [and] Life Sciences (in lieu of Aeromedical)."[7] In line with this proposal, the board Secretary, Colonel Gasser, prepared several alternate plans and sent them to Executive Committee members early in 1959. As Gasser later described it, these proposals were accorded a "dismal reception," with the commit-

---

*The Aerospace Weapons System Division, under the proposal, was divided into three areas of interest—vehicles, propulsion, and nuclear. The second Division was divided into two areas—"General Electrical Components," and "Sub Systems Areas." The first of these was further subdivided into four areas: (1) Pacing Components Science, (2) Sensing Techniques, (3) Communication Techniques, and (4) Analysis Techniques and Computer Logic.

See notes on page 188.

teemen now voting overwhelmingly to continue under the traditional organization.[8] Dr. Macdonald, for example, after concluding that the "present . . . setup is both sound and effective," expressed his belief that the SAB should strive for "more precise definition of functional areas and responsibilities, more effective use of liaison membership across panels, broader use of consultants, and, finally, continuation and perhaps extension of ad hoc committee activities."[9] Prof. Perkins noted that "regardless of panel structure, adequate technical coverage of the major fields of interest within the total membership . . . is essential . . . [and] major problems confronting the Air Force will require cross panel membership no matter how the basic panels are constructed."[10] Dr. Stever concurred, pointing out that "no matter what grouping of panels one selects one can always find some disadvantages and some advantages of the structure." In his view, the most important requisites of SAB organization were "flexibility and dynamic control" and the current organization possessed these to a satisfactory degree.[11]

Thus it was resolved that the membership buildup to meet the exigencies of the post-Sputnik era would be made within the traditional SAB organizational framework. As initial steps to the buildup, board officers introduced two important changes. In 1959 they established a new category of membership called "Senior Statesman" to which Drs. Draper, Kaplan, C. B. Millikan, Teller, and Wattendorf accepted appointment.* The reassignment of these long-time members opened their panel positions to "new blood" yet enabled the board to retain their services. When Dr. Millikan found the new assignment "somewhat mysterious," Dr. Doolittle offered the enlightening definition that "the Senior Statesman role is to be interpreted merely as a relief from tedious administrative duties with complete freedom to participate in board or panel activities wherever and whenever you feel it would profit the Air Force. In this way we hope to benefit from your abilities and time more than in the past."[12] In further recognition of their special value, Senior Statesmen also served on the Executive Committee.

The second change concerned the ex-officio membership category. From 1946 to 1951, the board had extended only one ex-officio membership—to the Director of Research and

---

*Dr. Getting accepted appointment as Senior Statesman in 1960 and Dr. Warren in 1961 (see Appendix E).

See notes on page 188.

Development, who also served as military director during those years. The position retained ex-officio membership when the military directorship passed to the Deputy Chief of Staff/ Development in 1952. Then, as noted earlier, Dr. Doolittle abolished the members-at-large category in 1954, offering many of the former incumbents ex-officio membership. The category was further broadened in 1956 when such memberships were extended to the Deputy Chief of Staff/Development, the ARDC commander, the Air Force Chief Scientist, and the Director of Development Planning (as well as the Director of Research and Development).

In the fall of 1959, General Putt notified General White that he questioned the wisdom of this practice, feeling that it violated the "philosophy of the SAB which is to be completely independent, objective, sometimes critical, and always observing from an outside perspective." The board had always "operated on the philosophy that civilian scientists from universities and industry bring an outside perspective to the Air Force," which ex-officio members might dilute since they "could not always be truly objective when faced with a divided loyalty."[13] Accordingly, Putt recommended abolition of ex-officio memberships, and General White concurred. The change was effected in September 1959 by amendment of the SAB regulation.

Essentially, then, the SAB had adjusted its policies so that added members would come aboard for just two reasons—to expand current panels and to staff new ones.* Prior to Sputnik, the board had generally followed the practice of limiting panels to six members or less, considering this to be the maximum figure for efficient operation. Now, however, panel chairmen were free to set manning more or less as circumstance dictated. As a result, all but one panel underwent a sizeable increase by 1962. The Electronics and Nuclear Panels grew to 10 members while the Geophysics Panel increased to 11 (including liaison members from other panels). Only the Reconnaissance Panel registered a decline, dropping from five to three members.† The rest of the increased membership staffed two

---

*A "members-at-large" category was reinstated at this time and remained in effect through 1962. Unlike the one which bore the same name and which Dr. Doolittle abolished in 1954, however, the new category was severely restricted in membership. It started at two in 1959, held at three for the next two years, and reached four in 1962, with one membership held by the Chief Scientist through these years.

†See Appendix E for a roster of membership, by panel, 1946-1964.

See notes on page 188.

new panels formed in 1959 and another established in 1961. Also in 1961, several experts in the general science of aerospace operations accepted board membership to staff a special permanent committee.

The first new panel descended from the Stever Committee on advanced weapons technology which had joined with Dr. Teller's Nuclear Panel to write the December 1957 space technology report. Thereafter, the committee had gone on standby pending further orders from General White on its future activities. Meanwhile, on Dr. Doolittle's request, the members kept themselves informed on Air Force space projects and considered possible board participation in the projects.[14] In mid-1958, the Executive Committee decided to form the committee into the Space Technology Panel whose role, as Doolittle explained to White, would be to "cover the entire spectrum from guided missiles through satellites and space platforms to manned interplanetary travel."[15]

General White concurred on July 24, 1958, and for the rest of the year the Executive Committee sought to obtain an acceptable panel charter. Since the panel would have to work closely with ARDC's Ballistic Missile Division (BMD), General Schriever was very much concerned with both the charter and the professional attachments of its members. Throughout 1958, Dr. E. H. Plesset served as chairman *pro tempore* pending clarification of the panel's duties. When Plesset accepted chairmanship of the Nuclear Panel, General Putt recommended that Dr. Stever chair the Space Technology Panel. Stever agreed and General Schriever expressed pleasure with the appointment.[16]

Initially, it appeared impossible to sketch out an area of activity for the panel which did not overlap or impinge upon the work of other SAB, Air Staff or ARDC agencies. Dr. Stever clarified the subject sufficiently by February 1959 through personal discussions with General Schriever and fellow SAB panel chairmen to enable his panel to at least begin to meet. Essentially his panel would concern itself with both ballistic missiles and space flight "from an overall standpoint" while, at the same time, other panels would still retain an interest in space from their own perspectives. To insure that the new panel did not intrude on other panel areas it would meet with them frequently "to take intensive looks at specific problem areas."[17]

_____

See notes on page 188.

As organized in early 1959, the Space Technology Panel had 12 members and seven consultants. It met several times in the winter and early spring, receiving briefings from USAF, Advanced Research Projects Agency, and National Aeronautics and Space Administration (NASA) representatives. Panel members also visited BMD, RAND, and several industrial aerospace plants working on space projects. In April 1959, General Putt sent the Air Staff the panel's listing of some 14 "problem areas in space technology" and asked for priority guidance for study.[18] The Air Staff provided the requested priorities but qualified them with the comment that "it is difficult to list the problem areas in space technology in order of importance as they are inter-dependent."[19] The Air Staff also noted that it was "limited in its areas of space research" by directives from the Department of Defense and that the SAB should "consider these restrictions" when recommending a course of action. To one SAB officer, the Air Staff reply seemed to intimate "that the SAB doesn't understand the problem."[20]

These Air Staff views reinforced what Dr. Stever and other panel members had already begun to suspect—that the panel was inadequately formed to serve the Air Force effectively on space matters. It was too large and too awkward to handle. Furthermore, there had been too many criticisms of "the industrial flavor" reflected in its membership.[21]

The Executive Committee discussed the problem at the fall 1959 board meeting and decided to reorganize the Space Technology Panel into a smaller unit and restrict it to SAB members only. It could then call on other panels when it needed expertise in component fields. Accordingly a new panel formed in late 1959 with eight members, of whom only Dr. Stever, who accepted the chairmanship, and Dr. Fred L. Whipple served it solely.[22] The other six—Mr. Edward J. Barlow, Dr. S. W. Herwald, Dr. John P. Marbarger, Mr. Perry W. Pratt, Dr. Radford, and Dr. Valley—held primary assignment on other panels and served in the capacity of liaison members. Dr. Stever, because of other commitments, had to relinquish the chairmanship in the spring of 1961 but retained panel membership. Prof. Perkins, recently returned to the board after serving as Assistant Secretary of the Air Force for Research and Development, became the panel head while continuing his regular duties as Aerospace Vehicles Panel chairman.

See notes on page 188.

During 1960, the reconstituted Space Technology Panel (1) met frequently for briefings on space projects and problems, (2) prepared one report on the need for closer Air Force/NASA cooperation on lunar exploration and another on the need for more vigorous action on space counterweapon sensor systems and missile target signature experiments, (3) provided verbal consultation on various space problems, and (4) assisted several SAB ad hoc space committees. At the close of the year, Dr. Stever submitted a summary estimate of the Air Force's space program to date. In the panel's opinion, the Air Force had about reached a point of diminishing return, technologically, in ballistic missilery. That is, the ICBM's already in operation or under development were about as effective as science could make them, and further increase in their reliability now rested with operations officers and site engineers. It was a different story with such newer aspects of the military space program as reconnaissance and warning satellites, satellite interception, and manned space vehicles. Here the surface of advanced technology had "only been scratched," and the panel suggested various means whereby the Air Force might proceed to exploit current knowledge in these areas.[23]

This proved to be the final report of the Space Technology Panel. The reorientation of the national space program in 1961 resolved most of the issues with which it had been concerned. Also, the nature of the space projects which the SAB undertook that year required a cross-panel approach. As a consequence, the Executive Committee concluded that the panel had outlived its usefulness and disbanded it in the fall of 1961.

The Air Force's increasing concern over the quality and management of its basic research facilities prompted establishment of a second new panel—Basic Research—in June 1959. Dr. Valley accepted the chairmanship and, by the fall of that year, had defined the panel's mission to encompass investigation of "all matters of policy, procedure and composition pertinent to Air Force basic research" and enhancement of USAF relations with the basic research community. Comprised of Dr. Valley and Dr. J. C. R. Licklider with Prof. Leo Goldberg and Dr. Charles H. Townes serving as liaison members, the panel met for the first time on December 7, 1959, and agreed that the Air Force, "to derive maximum benefit from new, advanced scientific knowledge," had to maintain its role as an active participant in the basic research community-at-large. The Air

See notes on page 188.

Force possessed a great number of outstanding scientists, the panel pointed out, but many were at work in areas where they were not particularly competent nor interested. To build and retain a good reputation in the scientific community, these persons had to be given greater freedom to work on projects more in keeping with their primary skills and enthusiasms. The panel members promised assistance in developing more satisfactory policies for in-house basic research and assignment of in-house scientists and for keeping the Air Force advised of instances when "this extremely valuable resource" was not properly utilized. The panel also promised to seek relief from the "many restrictions imposed by USAF and DOD budget, comptroller, and legal people which seemingly obstruct the optimum prosecution of research contracts."[24]

Soon after the Basic Research Panel formed, ARDC established the Air Force Research Division. Its commander, Brig. Gen. Benjamin G. Holzman, noted that now for the first time research had assumed "equal status with development and systems."*[25] The Basic Research Panel remained "intimately concerned" with the management and operation of the division through means of a special arrangement which permitted it to communicate directly with General Schriever. In February 1961, Dr. Valley complimented all concerned, noting that "enlightened plans and administrative changes" had fostered "a superior climate for research within the Air Force Research Division and . . . [increased] the Air Force's stature in the basic research community."[26]

In the fall of 1961, on Secretary of Defense Robert S. McNamara's direction that the services strengthen morale and standards in their in-house laboratories, General LeMay, Chief of Staff, called on the SAB to "examine research and development activities . . . with major emphasis on a drastic improvement of our in-house laboratories in accordance with the intent of [the directive]"†[27] Since the job was too vast for one panel,

---

*General Holzman held the distinction of being the only military officer ever to serve as a full SAB member while on active duty—as a colonel during 1946-1951.

†The directive stemmed from an inquiry from President Kennedy on whether contractors or direct government operations were most suited to conduct certain aspects of research. A study initiated by Dr. Harold Brown, Director of Defense Research and Engineering, and performed by the Weapons System Evaluation Group in response to the President's inquiry had uncovered many weaknesses in the laboratories.

See notes on page 188.

the SAB formed an ad hoc committee under Dr. Leonard S. Sheingold to carry out the study. About the same time, on General Schriever's request, the SAB formed another cross-panel committee under Mr. Richard E. Horner to review AFSC technical facilities. In February 1962, the latter committee recommended that research and development agencies be given direct control of certain sizeable research funds currently contained in the military construction budget.[28]

The Sheingold Committee report, submitted to General Le-May in April, endorsed this proposal as well as AFSC's plan to establish the Research and Technology Division (RTD).[29] RTD was duly activated in July 1962, and Maj. Gen. Marvin C. Demler, its first commander, later noted that as a result of the Sheingold Committee recommendations 37 small laboratories scattered across the country were consolidated into seven units under RTD. General Demler also attributed major credit for creation of the "Lab Director's Fund" to Dr. Valley because he had convinced Secretary McNamara's staff that establishing a single line item in the budget for each laboratory would both increase efficiency and save money. Demler stated that this action had "done more for the morale of people in the labs than anything since the 1958 Stever Report . . . [showing] that somebody really trusted the Lab Directors to do a good job with these unfettered funds."[30]

The computerizing of Air Force command and control systems which began in the late 1950's led to the formation of the third new SAB panel—the Information Processing Panel—in 1961. By 1960, SAB and RAND studies had pointed up the need to review the many government and industry proposals for the next generation of high speed, high capacity computers. In February, General Wilson asked the SAB to look at the area, giving particular attention to (1) the degree of cooperation and exchange of information among the many computer research groups, (2) the amount and appropriateness of effort being expended on bionics, and (3) the relationship of the characteristics of the next generation of high speed computers to those already in operation or under development.[31]

SAB's Electronics Panel responded to the request, submitting its report in April 1960. This finding, which General Wilson called "an outstanding contribution . . . of great assistance to the Air Force," pointed up "vividly" what the Air Force

See notes on page 189.

suspected: recent advances in technology had created an urgent need for greater research and experimentation in the overall communication field.[32]

Accordingly, General Wilson—who as military director summarized annually in a report to the SAB chairman the year's gains and suggested possible areas of study for the coming year—proposed command and control as one of the areas most in need of SAB attention in 1961. In Wilson's eyes, the Air Force required a control of its forces "so positive and sure that the chances become nearly infinitely remote that a third World War could be started inadvertently either by the USSR or by this country."[33] Further stimulus for a greater SAB effort in this area came during the October 1960 general board meeting at Hanscom Field, Mass., when SAB members examined the complexity and promise of computerized command and control systems in detail.

Following this meeting, the SAB Executive Committee appointed Drs. Launor F. Carter, Radford, and Valley (chairmen of the Psychology and Social Sciences, Electronics, and Basic Research Panels, respectively), and Mr. Schenk, consultant to the Electronics Panel, to explore the advisability of creating a panel in the area. After meeting with Brig. Gen. Baskin R. Lawrence, commander of AFSC's Electronic Systems Division (ESD), at Hanscom in March 1961, the group voted in favor of the move. Whereas the Electronics Panel would concern itself primarily with techniques and system design, the new Information Processing Panel would deal primarily with function. Specifically, it would study such matters as (1) decision making and command, (2) information retrieval, (3) operational integration of the several systems currently under development, and (4) application of new techniques for computer programming or apparatus construction to information processing.

The Executive Committee concurred in the plan and, shortly after, Dr. Walter A. Rosenblith accepted the chairmanship. By December 1961, the new panel had formed with Mr. John D. Madden, General Earle E. Partridge USAF (Ret.), Dr. Frederic M. Tonge, and Dr. Willis H. Ware as members and Dr. Burton F. Miller as liaison with the Electronics Panel. When Rosenblith found it necessary to resign board membership in 1962 because of the press of his university duties, Dr. Gerald P. Dinneen transferred from the Electronics Panel as his replacement.[34]

See notes on page 189.

A fourth new group which eventually was accorded panel status evolved from SAB actions initiated in 1959 to assist the Air Force in the formulation of arms control policies. In January 1960, a SAB ad hoc committee formed the previous year under Dr. Doolittle presented its views to General White on the effect the atomic test moratorium would have on the Air Force mission.[35]   The report, which White called "a great job in handling a controversial subject with great skill," subsequently formed the basis for the Air Force position on arms control negotiations.[36]   Later that year, White accepted SAB's offer to create a standing committee that would keep active watch on the subject to provide the Air Force "with important and highly useful evidence and (hopefully) plans on which to implement action should international negotiations fail to materialize in a vein acceptable to the United States."[37]

As originally formed in early 1961, this Arms Control Committee consisted of Dr. Overhage, chairman, and 11 members, of whom about half held primary assignment to the committee and the rest served as liaison from the panels. In June 1961, Overhage, because of the demands of other duties, turned the chairmanship over to Dr. Carter but remained on the committee. Soon after, Prof. Thomas C. Schelling accepted the post when Carter became Chief Scientist of the Air Force. Meanwhile, the committee had isolated as its area of interest, as Schelling expressed it, everything "designed to reduce the danger of unauthorized action, of false alarm and misinterpretation, [and] of communication failures."[38]

See notes on page 189.

CHAPTER TEN

INTIMATE AID ON RISKY DECISIONS

> AFSC is the largest governmental
> agency in the systems acquisition
> business, which encompasses broad
> areas of research and engineering.
> As such, the Command needs the
> best advice and counsel that can
> be obtained. [Since] much of the
> decision making and detailed plan-
> ning occurs at the AFSC Division
> level . . . advice [is needed] at that
> level from the scientific commu-
> nity, as embodied in the SAB.
>
> —Bernard A. Schriever*

Between 1954 (when the board expedited the dispatch of its reports to ARDC) and 1959, little was accomplished in the way of strengthening the SAB-ARDC relationship. The 1957 Perkins Report had applauded instances where SAB members had served as consultants to ARDC centers and recommended an expansion of this activity. But, as noted earlier, circumstance thwarted follow-through on the report and SAB-ARDC relations continued on the same desultory course in ensuing years.

Portents of improvement first appeared in the spring of 1959. Ex-ARDC commander Putt had become SAB chairman, and the newly-appointed ARDC commander, General Schriever, had initiated the organizational overhaul that eventually produced the AFSC. Both agreed that ARDC field units required more direct counsel of the sort that SAB members could give. "SAB assistance to ARDC has been a longstanding subject, both with the board and myself," Putt declared, and "the cli-

---

*In remarks at AFSC/SAB meeting, Headquarters AFSC, April 2, 1962.

mate in ARDC . . . [now appears] conducive to working out the mechanics."[1]

As a start, the board and ARDC set out to bring SAB services to bear on major technical obstacles in current ARDC applied research and development programs.[2] ARDC listed their problems as they arose and sent them to SAB for counsel. The board then assigned them to panels or formed ad hoc groups if they cut across the several disciplines.[3] In this manner, as General Wilson viewed it, the board sought to help shoulder the "many difficult and perhaps risky decisions" General Schriever was obliged to make during these critical times.[4]

However, General Schriever still required more help from science than SAB was providing him, even through the improved procedure. "The SAB has performed an invaluable service in the past and has certainly brought the civilian scientific community closer to the Air Force," he wrote General Putt in September 1959. But SAB remained geared primarily to serve the Chief of Staff and, hence, was not able to afford ARDC the intimate service it needed. For a time, Schriever considered enlisting the assistance of the National Academy of Sciences, noting that the Army and Navy had availed themselves of this service by supporting permanent committees of the National Research Council and that the Air Force, through ARDC, could profit from a similar arrangement. He felt there would be more than enough critical ARDC scientific problems to keep both the SAB and an Academy advisory committee gainfully employed.[5] However, he subsequently dropped the project, informing Putt in October 1960 that "the establishment of an ARDC Advisory Board at this time is no longer being contemplated."[6] The SAB thus remained ARDC's primary scientific advisory source. At the same time, in a radically different approach to SAB-ARDC relations, it also, in future years, succeeded in giving General Schriever more of the personal, intimate aid he required.

The precise beginnings of the new venture—which became known as the SAB-ARDC Division Advisory Group (DAG) program—defy documentation. Essentially, as Colonel Gasser pointed out, it evolved out of the "determined attempt by General Putt to satisfy General Schriever's desire to have greater access to the SAB and still remain within the board's charter of reporting only to the Chief of Staff."[7] In any event, Putt and Schriever had worked out the idea sufficiently to present it

See notes on page 189.

to the SAB Executive Committee in January 1961. About the same time, Putt solicited and received General White's oral concurrence. Finally, SAB members and ARDC officers forged it into a working concept at a March 1961 meeting.[8] in effect, therefore, the birth of the new venture coincided with that of the Systems Command.

Dr. Stever briefed General White on its basics in April 1961. The board would set up "selected SAB groups" to determine directly with AFSC division commanders the nature of their problems and provide appropriate advice after suitable investigation and deliberation. This would be done in an informal and confidential manner. In keeping with the last objective, Stever noted that reports prepared by the groups would go only to the AFSC division commanders.[9]

The SAB Executive Committee approved the project at an April 1961 meeting, agreeing to provide "Division Advisory Committees [of] appropriate scientific talent to AFSC Commanders." Dr. Stever undertook to prepare formal terms of reference defining the relationship between the SAB groups and division commanders;[10] however, at a subsequent meeting, Putt and Stever explained that the board and AFSC would have to work out many of the details as they went along. The SAB's whole object was "to assist AFSC in every way practicable via [the] Division Advisory Committees and any other special groups or bodies needed." For the time being, they said, "this particular SAB approach to assisting AFSC would . . . have to be an experimental one inasmuch as the actual procedural actions to be followed cannot be frozen at this time."[11]

In May 1961, SAB and AFSC settled on the name—Division Advisory Group. Also in that month, the first DAG—organized in April with Dr. Radford as chairman to assist ESD—held its first meeting.[12] In June, the Executive Committee approved procedures for creating and operating additional DAG's. The SAB furnished a list of SAB members and consultants to each AFSC division from which the commander selected those advisors that appeared most qualified to assist him. Every effort was then made to list advisors most conveniently located, geographically, to the division. This reduced travel burdens and enhanced availability of the advisors. The fact that the DAG's were SAB-sponsored did not rule out the employment of specially qualified individuals who were not SAB members or consultants so long as AFSC assumed responsibility for such matters as the granting of security clearance. On operations, the Executive Committee again stressed the point that while DAG membership would be composed primarily of SAB members, the DAG's themselves would not constitute a SAB activity. The advisors would be viewed simply as "scientific experts [made] available [to AFSC] from the SAB roster." However, the Executive Committee recommended that administrative responsibility (per diem, travel, fees, security, etc.) for SAB members serving with the DAG's remain with the SAB secretariat.[13]

The second DAG formed in December 1961 for the Space Systems Division (SSD) under Dr. C. B. Millikan's chairmanship.[14] The third, for the Ballistic Systems Division (BSD), met for the first time in March 1962 with Dr. Teller as chairman.[15] The next month, SAB officers and DAG participants joined General Schriever and the division commanders at AFSC Headquarters to assess DAG experience to date and reaffirm basic principles of the undertaking preparatory to proceeding with the final DAG activations.

At this April review, Maj. Gen. Charles H. Terhune, ESD commander, reported that the ESD DAG had found that "too frequent formal meetings were not good because so many of the problems were discussed on an informal basis." After each meeting, this DAG had prepared an informal report which it distributed on a very restrictive basis. Dr. Millikan and Maj. Gen. Osmond J. Ritland, SSD commander, said the SSD DAG had followed similar practices, adding that they also advocated an informal relationship between commander and DAG chairman. Dr. Albert L. Latter, vice chairman of the BSD DAG, concurred, indicating his DAG's experience that "a forthright and intimate exchange of views" had to prevail on all matters if the advisors were to be fully effective. General Schriever "was very forceful" in support of this philosophy of DAG operation. Complete intellectual honesty, and intent was mandatory to successful DAG operation, he said, which meant that a DAG "should have available all prior decisions relative to the problem and all prior information bearing on the problem," with nothing withheld.[16] Confirmation of this key point—by experience as well as by intent—thus eliminated the one significant objection ever raised in SAB council concerning the DAG venture. When the idea was first presented in early 1961, the Executive Committee feared that the groups might be em-

See notes on page 189.

ployed to accomplish staff work rather than to assist in scientific and technical areas. The April meeting made clear that such had not happened and that General Schriever intended to personally prevent its happening in the future.

With everyone now in full agreement on the validity and worth of the program, the board acted to form the remaining DAG's. Establishment of the Aeronautical Systems Division DAG came later than originally planned. Prof. Perkins had agreed to chair it, but the increased duties that came his way following his acceptance of the SAB vice chairmanship in early 1962 eventually forced his withdrawal.* In April 1962, Dr. Alexander H. Flax accepted the chairmanship and this fourth DAG formed in May and met for the first time in June.[17]

In July 1961, after AFSC's Atlantic Missile Range (AMR) had received the job of coordinating all Defense Department range resources in support of NASA's manned lunar program, the SAB, on General Schriever's request, had formed an ad hoc committee under Dr. Brian O'Brien to help with the expanded program.[18] Following an August 1961 meeting with AMR officers, the Committee had recommended creation of a Range Technical Advisory Group (RTAG) that would respond to range instrumentation problems of all AFSC divisions.[19] General Putt and General Schriever concurred and, by May 1962, this fifth DAG had formed with Dr. Sheingold as chairman.[20] It met for the first time in June to assist in the job of creating, as one AFSC officer expressed it, "a cohesive and efficient management system for range instrumentation on a global scale."[21]

General Schriever approved the charter for the sixth, and final, advisory group in July 1962, and it met for the first time in October. This was the Foreign Technology Division DAG, under Dr. O'Brien's chairmanship, which would help to obtain and assess foreign technical information.[22]

DAG chairmen and division commanders exercised a nearly free hand in appointing members and consultants to the DAG's through 1962. As a result, DAG's differed considerably in size and composition. For example, by September Dr. Teller's BSD DAG had nine members, of whom six were SAB members; Dr. Millikan's SSD DAG had 13 members, seven of whom held SAB membership.†

---

*See Chapter Eleven, page 111.

†See Appendix G for a roster of SAB/DAG members, 1961-1964.

See notes on pages 189 and 190.

The initial division regulations on the DAG's were equally disparate, but General Schriever saw no advantage here. Consequently, in the spring of 1962 he assigned the AFSC staff the job of aligning the operation, including division regulations, into a reasonably uniform pattern. ESD Regulation 20-2 of September 1962 exemplified the results of these labors as well as illustrated how the initial concepts of operation had evolved. The directive noted that DAG members and consultants would be selected from active SAB lists or, as necessary, from outside the SAB. The SAB chairman and division commander would jointly select group members on an annual basis. The division commander would create a secretariat to service the DAG for such administrative duties as informing the SAB secretariat and AFSC Headquarters of meetings, preparing meeting agendas, keeping DAG files, and processing reports. From a management standpoint, the DAG would be responsible to and report directly to the division commander and its findings and recommendations would not carry SAB endorsement unless· they were subsequently acted on by the board in its official capacity as advisor to the Chief of Staff.[23]

See notes on page 190.

PART IV

NEW CARROTS FOR WINGED HORSES

Definitely a boring recital of all the obvious should be avoided. The primary object is to put new carrots in front of the noses of the Air Force winged horses.

——*H. Guyford Stever*, in letter to Dr. S. W. Herwald, December 21, 1960.

Probably the most important single element of the scientific strength of the Air Force outside of its own organizational structure is the Scientific Advisory Board. This brings to bear on Air Force problems some of the best scientific and technical talent in the country. . . . I completely agree with the . . . statement by Courtland D. Perkins, recently the Assistant Secretary of the Air Force for Research and Development: "I think that of all the scientific advisory committees that I have ever had any dealings with the Air Force Scientific Advisory Board is used the most effectively." This does not obviate the possibility of improvement, or the need for changing patterns of operations to meet changing needs. In making improvements and changes however, care must be taken that what is already very good is not destroyed or disrupted.

——*Alexander H. Flax,* in memo to General Curtis E. LeMay, March 15, 1961.

CHAPTER ELEVEN

HAZARDS OF THE BUSINESS

> When we started this board, I had
> a long talk with General Spaatz ...
> and also wrote him a letter saying
> "It is now peacetime, and it is nec-
> essary that, in order to have expert
> opinion and talent, we not consider
> the individual member's connec-
> tions with a firm, scientific society,
> or with a university because these
> organizations have contracts with
> the Air Force to develop some-
> thing." General Spaatz answered it
> is clear to the Air Force that the
> position of the Advisory Board
> members should be above this
> point of view.
>
> ——Theodore von Karman*

From its inception, the SAB had enjoyed complete freedom in choosing members and determining its total membership requirements. Officially, final decision on appointments rested with the Chief of Staff, but there is no indication that he ever refused the recommendations of the board chairman and military director. The situation changed in 1962, however; new and more stringent controls over the use of civilian advisors introduced by the Kennedy Administration deprived SAB of much of its erstwhile freedom in these matters. And application of the controls necessitated significant changes in other aspects of board procedures and operations.

Prior to 1962, SAB philosophy on preventing conflict-of-interest in its membership remained direct and uncomplicated. Recognizing that members would be privy to information which, conceivably, they could put to personal gain, board founders at the 1946 charter meeting adopted the policy that "other

*In speaking to the SAB Executive Committee, March 18, 1948.

things being equal, it [was] preferable to appoint members from universities or other non-profit organizations rather than from industry." This did not automatically solve the problem, of course; the qualities which attracted the SAB to a university person also made him desirable to industry, and most of the charter members served one or more private firms as consultants. But Dr. von Karman and General LeMay had no fears on this count, noting they did not think "the integrity of any of this group will ever be questioned." In other words, such men were *a priori* above reproach.[1]

The assumption was sound, since the board adhered to it over the following years without repercussion. Meanwhile, beginning with the board buildup during the Korean War, board officers found it necessary to make many exceptions to this basic policy. As Dr. Root noted in 1952, his Aircraft Panel had "drawn from the relatively few universities and non-competitive corporations to the point where it [was] now quite difficult to make suggestions for new members."[2] By 1953, board membership included 13 persons whose primary work affiliation was with a competitive, industrial organization. In each case, it was pointed out, "the appointment [had] been made with full recognition of this affiliation and with the belief that the individual concerned was eminently qualified and completely ethical."[3] The question was now raised whether the practice should be continued, and the Executive Committee replied in the affirmative. By 1954, there were enough industrial members to prompt board officers to report the fact to the Chief of Staff and Secretary of the Air Force so that they might be prepared if anyone questioned the matter. In explaining the situation, Dr. Kelly commented that "recriminations were hazards of the business" and the best the board could do was "strive to keep [its] academic proportions high" without sacrificing quality of membership.[4] In other words, while the board did not want to appoint members from competitive industry unnecessarily, it also wanted the best man in the field no matter his affiliation.

To protect the Air Force and itself from criticism on its increasing industrial membership, the board applied ever more stringent, self-imposed controls. For example, the board's military director, General Putt, after ARDC inquired into the handling of presentations by one aircraft firm when SAB members from competitive firms were present, advised that command

See notes on page 190.

not to volunteer any presentation without permission from the firm concerned and, when possible, to let the firm give it. If the problem were especially serious, Putt advised ARDC simply to inform the SAB secretary that the presentation should not be given.[5] Also, in at least one instance the board chairman and a panel chairman agreed that the nature of the current work of that panel made it necessary to reassign its industrial members to other board positions.[6]

After the Kennedy Administration came to office, Secretary of Defense McNamara served notice that he intended to take a more personal interest than his predecessors in the numbers and types of advisory committees. On March 23, 1961, Mr. Roswell Gilpatric, Deputy Secretary of Defense, instructed all agencies to coordinate with his office on the "appointments of candidates selected as consultants and experts in the Departmental service, other than medical, who will be retained for more than 30 working days within a year." He asked for a brief biography of each advisor and an account of the service provided. The procedure applied to renewals as well as new appointments.[7] Again, in July 1961, General LeMay furnished Dr. Harold Brown, Director of Defense Research and Engineering, a detailed description of SAB composition and mission in response to Brown's request for data on which to assess the value of current advisory committees.[8] When no further action ensued on these matters in the months immediately following, the SAB assumed it had weathered scrutiny. Before the year was out, however, the SAB became caught in a maelstrom that threatened to topple it from its traditional foundations.

In December 1961, Mr. John W. Finney of the *New York Times* wrote several articles questioning the propriety of General Putt's position as SAB chairman. He pointed out that Putt, after retirement from the Air Force, had accepted the presidency of the United Technology Corporation to which the Air Force had subsequently awarded a $2 million research contract. While he did not accuse anyone of wrong-doing, Finney found it curious that the Defense Department would approve Putt's advisory appointment but that the Atomic Energy Commission, under its procedures, would have declared him ineligible. In other words, the two agencies seemed to hold to different standards with AEC's appearing the stricter.[9]

Mr. Finney's articles did not influence the appointment of SAB officers for the coming year. As early as October 1961,

General Putt had informed General LeMay that he had found it increasingly difficult from his West Coast location to give SAB business sufficient time and attention and, for this reason, would decline reappointment. Consequently, General Le-May offered the chairmanship to Dr. Stever and the vice chairmanship to Prof. Perkins for 1962, and they accepted and prepared to assume their new duties. However, late in December 1961, Mr. McNamara asked Secretary of the Air Force Eugene M. Zuckert to hold up all SAB appointments "pending resolution of conflict-of-interest considerations [on DOD] advisors and consultants."[10] Reluctantly, Mr. Zuckert issued the necessary orders. At the same time, he and Mr. Max Golden, Air Force General Counsel, agreed that the circumstances of the situation justified their asking McNamara to "place the entire issue in a DOD context rather than singling out the SAB . . . or [SAB] individuals."[11] Meanwhile, to provide the board with interim leadership Dr. Stever and the board's new military director, Lt. Gen. James Ferguson, agreed that between General Putt's resignation and official confirmation of Stever's appointment, Stever would "act as if he were Chairman for the purpose of carrying out important . . . functions of the Board."[*12]

On request of President Kennedy, the Attorney General reviewed the entire subject of government use of advisory committees, submitting an opinion on January 31, 1962.[13] This was followed on February 26 by a Presidential executive order setting forth a new set of procedures standardizing the appointment and employment of advisors throughout the government.[14] Secretary McNamara on March 12 issued an implementing directive (DOD Directive 5500.8) and on April 24 the Air Force followed suit (AFL 40-15).

Under the new procedures, all agencies would keep a "Statement of Employment and Financial Interests," updated semiannually, on each advisor, to include the names of all companies or research institutions in which he and his immediate family held securities or other financial interest. The SAB secretariat solicited the statements from members in April 1962 and then had appropriate Air Force offices review them for possible conflicting interests.[15] Upon completion of the review, the statements were returned to the secretariat. On April 17, the Assistant Secretary of the Air Force for Research

*General Ferguson succeeded General Wilson as Deputy Chief of Staff/ Research and Technology and as SAB military director in the fall of 1961.
See notes on page 190.

and Development, Dr. Joseph V. Charyk, forwarded the statements of Dr. Stever and Prof. Perkins to Mr. Gilpatric, who
promptly approved the appointments.[16]

The new chairmen endorsed the remainder of the SAB
secretariat's program for applying the new standards to SAB
operations. These included such precautions as assuring the
presence of a full-time SAB officer at all SAB-sponsored meetings for (1) adjourning any meeting when this action seemed
to be in the public interest, (2) certifying in writing as to the
correctness of meeting minutes, (3) keeping verbatim minutes
of Executive Committee and general board meetings and of
meetings at which the majority of attendees were from industry, and (4) keeping a record of the affiliations of non-
members attending SAB meetings. This officer also insured
that industrially affiliated members did not participate in any
proceeding in which their firms were directly involved. Finally,
the new procedure required the secretariat to inform the military director of study requests before panels or special committees acted on them.[17]

During the standards of conduct review, Secretary Mc-
Namara apparently noticed for the first time the considerable
difference in size between the SAB and its counterparts in the
other services and suggested to Secretary Zuckert on June 21,
1962, that SAB trim its roster to 20 members. Air Force
leaders and SAB officers met immediately to prepare a defense
of the current board. Past experience provided ample evidence on why the board could not function in its traditional
manner with such a small membership. However, experience
also showed that it had operated quite effectively in the past
with fewer members than were currently on the roster. General LeMay, acting on recommendations of SAB officers, proposed a compromise on July 3 which Mr. Zuckert passed to Mr.
McNamara two weeks later. The board would fix its maximum
membership at 70. To satisfy McNamara's basic desire—that
the office of the Secretary of Defense be kept current on advisory appointments and advisory committee actions—Zuckert
noted that henceforth the SAB would submit its studies directly
to the Air Force Secretary as well as to the Chief of Staff. In
other words, the SAB would be advisors to both the USAF
civilian and military leadership.[18]

Mr. McNamara agreed to these terms and Mr. Zuckert and
General LeMay formally approved them in August 1962, to be

See notes on page 190.

effective January 1, 1963.[19] On November 23, 1962, a revised edition of the SAB regulation made the changes official with the key revisions reading as follows (italics added) :

> The composition of the entire Board is subject to periodic review and approval by *the Secretary of the Air Force* and the Chief of Staff. The chairman transmits all findings and recommendations to *the Secretary of the Air Force* and to the Chief of Staff. . . . Normally on the recommendation of the Chairman, *and with the approval of the Secretary of the Air Force*, the Chief of Staff appoints members and determines the length of their appointment.

Meanwhile, the SAB Executive Committee had met in special session on August 4, 1962, to begin the task of deciding where and how to cut back.[20] The board currently numbered 88 members, which meant board officers had to rotate 18 to comply with the "70-max" ceiling. As a start, they disestablished the Reconnaissance Panel, rotating two members and reassigning the third. The action was advisable, Dr. Stever noted, since the panel had been dormant for the past year.[21] They then realigned the Arms Control Committee into a panel, reducing its membership to three.* Finally, they established a new category called Associate Advisors and invited the several eminent scientists on the board who were in government employ to serve in this capacity. The board would continue to invite them to board meetings and to serve on panels and special committees, but would not count them against the membership ceiling. This change was actually a long overdue clarification; as government employees, these members were not advisors as stipulated in standards of conduct regulations. Neither were they consultants in the sense that they contributed only when their services were required on specific projects. SAB regarded them as top figures in their fields, desired their services regularly, and, because there had been no reason to the contrary, had dignified their services with full board memberships.[22]

The Executive Committee levied the remainder of the reduction on the panels and entered 1963 with a membership of 70. Later, the roster dropped to 68 when Chairman Emeritus von Karman died and Senior Statesman Warren resigned. Membership remained at this figure through 1964.

Meanwhile, the requirement to reduce membership had forced the Executive Committee to reassess the board's DAG

---

*Board officers also disbanded the small members-at-large category at this time, reassigning persons holding this assignment to panels.

See notes on pages 190 and 191.

commitments. General Ferguson and Dr. Stever conveyed the committee's conclusions to General Schriever in September 1962, notifying him that the board felt that it had to reduce each DAG to not more than five SAB members.[23]   At the same time, the SAB indicated its willingness for AFSC division commanders, whenever they felt a DAG did not "possess all the talents required on a specific problem area," to add consultants from outside the SAB membership.[24]   General Schriever notified his commanders of the change in November, forwarding to them the reduced DAG rosters which the board had drawn up. The use of special advisors on DAG's, Schriever noted, was intended in no way "to circumvent the reduction of the DAG to a five-member group," and the advisors would be employed "only on a temporary or intermittent basis, to accommodate special or unique requirements."[25]   Accordingly, SAB membership on each of the six DAG's was reduced to five at the beginning of 1963.

In February 1963, Dr. Stever notified General Schriever that the SAB could support an AFSC recommendation that SAB members employed by non-profit organizations be permitted to take part in DAG operations as long as they were carried as DAG associate members and not charged against the SAB's DAG membership quotas. Stever agreed that such an arrangement promised many benefits. Members from non-profit organizations were in a position to give more time to Air Force matters than their colleagues from the universities and industry. Also, the AFSC divisions and the non-profit organizations which supported them worked so closely that both profited by the work of the DAG's.*   Finally, as General Ferguson pointed out, use of these advisors would afford the DAG's an invaluable additional source of scientific and technical talent. The board issued an official policy statement on the arrangement in April 1963.[26]

The SAB Executive Committee reviewed the revised DAG program in the spring of 1963, expressed pleasure at its overall effectiveness, but noted several problem areas. DAG activities sometimes overlapped panel activities; on a few occasions DAG's had been asked to work on "nut and bolt" matters which

---

*Adequate safeguards were established to insure that members of non-profit organizations serving as DAG associate members were not placed in the position of reviewing or passing on their companies' activities and contractual arrangements.

See notes on page 191.

division staffs could have handled just as easily; and, finally, division DAG offices had not always kept the SAB secretariat promptly informed of scheduled meetings.[27]   General Ferguson passed these criticisms to General Schriever who called them to the attention of his division commanders.[28]

From this point, the DAG's proceeded without further alteration or difficulty and, from all reports, in a highly satisfactory and effective manner. In January 1964, General Schriever noted his division commanders' reports that the DAG's had "been most helpful in the definition and solution" of their problems."[29] Secretary Zuckert confirmed the assessment the following August, stating that "although reports of the DAG's do not usually come to my attention, I am informed that [they] are providing significant assistance to their respective AFSC Divisions and that there is general satisfaction with the results."[30]

See notes on page 191.

CHAPTER TWELVE

KEEPING COMPETENCE INTACT

> The big problem now, in my opinion, is being able to cope with the new weapon systems that seem to be coming along so fast that they tend to overwhelm management; . . . it is just this very time when the military services must retain a great nucleus of scientific strength to provide continuity . . . [and] maintaining the quality of the SAB is a necessity in order to keep this competence intact for the future.
>
> ——Eugene M. Zuckert*

The restraint placed on SAB during the revision of the conflict-of-interest regulations in 1962, the cutback in SAB membership in 1963, and the new criteria set by the Secretary of Defense for system and advanced development studies created problems for Dr. Stever and General Ferguson, the new SAB chairman and military director, and the other board officers which had little precedent in board history. Though the DAG operation survived these changes and in General Ferguson's words, "resulted in accelerated participation by many members of the SAB on programs and problems of immediate concern," the more traditional SAB activities registered an immediate decline and, in 1963 and 1964, dropped sharply. The figures on SAB reports and meetings between 1959 and 1964 afforded some insight into the extent of this decline :†[1]

---

*In statement to SAB Steering Committee at a luncheon meeting on January 7, 1964.

†As reported in the SAB annual reports. These yearly reports from the SAB chairman to the Chief of Staff were introduced in 1959 and contain, in addition to the information cited, brief accounts on studies submitted, copies of significant replies received on the studies, and resumes of Air Force consultative services performed by SAB members in addition to their regular board duties.

See notes on page 191.

	1959	1960	1961	1962	1963	1964
No. of Formal Reports	20	19	21	17	13	15
No. of Special Memos	12	10	29	12	8	8
No. of Panel/Ad Hoc Meetings	48	59	86	62	40	40

For the Air Force and SAB, the situation which began in 1961 was analogous to that immediately following World War II in the sense that final decision on most key USAF requirements again rested outside the Air Force. As a consequence, Air Force research and development project requests, and SAB contributions to them, had to be more comprehensive and definitive than ever before. Reports submitted to the Chief of Staff by Dr. Flax, while serving as Air Force Chief Scientist, and a SAB committee in 1961 pointed up the complexity of the new situation.

The first report was prepared by Dr. Flax in response to General LeMay's request that he investigate why so many Air Force programs failed to achieve "complete and unqualified approval" at higher scientific levels.* Replying in March 1961, Flax traced many of these failures "to considerations which are not purely technical or scientific but which are by legal or administrative assignment or by public acceptance within the purview of these [higher echelon] groups." He noted that the "other considerations" included such matters as "the validity of military requirements, operational concepts, total costs and cost-effectiveness relative to systems proposed by other services, survivability, penetration capability, and even political factors such as degree of stability of deterrent and degree of provocation involved in employment."[2]

In a similar request to the SAB, General White asked for an assessment of the Air Force's effectiveness in its research and development relations with the Department of Defense, the President's Scientific Advisory Council, Congress, and other higher agencies, as well as counsel on the changes the Air Force should make in policy, organization, and procedures to obtain better results in its dealings with these groups. White noted that the Air Force had a "broad and varied array of scientific talent available" which had made "important contributions to the nation and the Air Force." He felt, however, that much more could be done, that with such a plenitude of expert advice

---

*Vice Chief of Staff at this time, General LeMay, as noted in the preceeding chapters succeeded General White as Chief of Staff in June, 1961.

See notes on page 191.

"the Air Force should be able to produce better plans, requirements, and programs, and . . . should have far less difficulty in getting our point of view across to the other interested agencies of government."[3]

Dr. Stever accepted chairmanship of the seven-man committee on scientific resources formed to conduct the study and, in March 1961, met with General White and confirmed the fact that the basic objective was to counsel the Air Force on how to do "a better job of integrating scientific advice into [its] thinking."[4] Then, beginning in early April, the committee held a series of meetings with leaders of principal contractor laboratories and top Air Force officials, including Dr. Flax, General LeMay, and General Schriever, and submitted a report on May 26. The members felt that the Air Force had been "reasonably successful, in fact very successful relative to other services, in presenting its views and getting approval of forward-looking programs" from higher echelons. But the situation could be improved greatly by "producing plans, requirements, and the programs that are more acceptable scientifically and judged to be of top priority by the future-oriented, technologically capable administrators in the higher echelons of government." They described this new breed of administrators as[5]

> . . . technically trained and experienced [persons who] have increased the emphasis to be placed on the technical characteristics of weapons. They have also learned that the technical aspects cannot be separated from the economic, political, military, etc., so they consider all aspects of the problem. They are not reluctant to analyze the military desirability of weapons, plans, programs, and requirements; though they recognize that, in most cases, they have not had military operational experience, they believe that a combination of scientists' logical approach to problems and long experience with military men and military affairs, permit them to make reasonable judgments. Still another characteristic which must be taken into account is that they believe in going directly to individuals and groups who can speak with authority on a subject; they go directly to the working laboratory or industry that originated a technical idea for information and advice on the technical programs. This often short circuits the sometimes unwieldly information channels which they find within the Air Force.

The committee charted organizational, procedural, and conceptual revisions the Air Force should make within its research and development process if it were to satisfy the new standards set by the new administrators. Air Force agencies should become more familiar with "in-house" scientific re-

---

See notes on page 191.

sources (including RAND, the Aerospace Corporation, and the SAB), and call on them for aid in preparing requirements, plans, and programs. There should be direct and more frequent contact among these scientific resource groups, with a new top-echelon Air Staff agency responsible for "achieving a final concert of opinion for decisions on important technical systems." Other proposed measures included more technical education for Air Force officers, greater use of the Chief Scientist, more emphasis on Air Force basic and applied research programs, and a broader use of the SAB.[6]

Meanwhile, the SAB had taken a look at its own record and concluded it, too, needed to take a more far-ranging and comprehensive view in its future work. As the board secretary summarized the discussion of this point:*[7]

> Observation of board operations . . . has brought to light the fact that a sizeable percentage of board activity has been only partially productive. There appear to be a number of reasons for this situation. However, two seemingly outstanding reasons are: (a) the lack of a thorough understanding by the board of a number of the problems as posed, of their background, and of interface problems; (b) a seeming lack of appreciation of the need for optimum integration of scientific-technological findings and any recommendations arrived at with both short and long term aerospace operational requirements affected thereby. Difficulties encountered have been manifested in terms of at least partially erroneous interpretation by the board of the problem received due to either brevity, generalities and/or poor definition. This lack of clarity has, in a number of instances, led to panel or committee uncertainty as to its actual charge. It has also given rise to investigation procedure that has failed to take into account all available significant inputs thereby causing findings and recommendations to be something less than the high caliber and useful nature that they should have been.

Several members made differing proposals during 1961 on how the board might "modernize" its structure and procedures to overcome the objections noted above. However, the press of other affairs forestalled action on the subject during that year. The same held true in 1962, with one important exception. After Dr. Stever became chairman, he and General Ferguson promptly concluded that the board had "become a receptacle for any and all kinds of problems and, consequently, too many problems."[8] Accordingly, after coordinating their solution to this problem with the Executive Committee, they

---

*Colonel Gasser, secretary at this time, was succeeded by Colonel Robert J. Burger in June 1963.

See notes on page 191.

formed the SAB Steering Committee. Consisting of the board chairman, vice chairman, and military director, the Assistant Secretary of the Air Force for Research and Development, and the Air Force Chief Scientist, the Steering Committee met for the first time in February 1962 and, henceforth, became the initial recipient of all matters presented to the board for investigation. As Dr. Stever later explained the new group's purpose to General LeMay:

> This Steering Committee has as its objectives the monitoring of the incoming requests for board activities so that ad hoc committees of optimum effectiveness can be established, so that the appropriate resources of the Air Force can be brought to bear on SAB activities and so that the SAB reports can be made available and briefed to agencies, both within and outside the Air Force, that can effectively use the information to help the Air Force.

Creation of the Steering Committee, he concluded, was a preliminary move to make "more efficient use of SAB resources."[9] Subsequently, the Steering and Executive Committees, in concert, extended this drive to bring steadier and more aggressive guidance to board affairs by trimming Executive Committee membership and giving it "more responsibility for substantive rather than administrative matters."[10]

In early 1963, with the chore of reducing board membership out of the way, SAB officers tackled in earnest the long-deferred question of, as Dr. Valley expressed it, "whither the board."[11] In one statement of the essential nature of the question, Dr. Stever explained that with the DAG's taking over the "close contact and work on specific technical problems which formerly had been accomplished by the SAB," the board officers now desired "to look for a new broader mission for the SAB."[12] The search which continued over the greater part of the next two years, opened the door wide to critical appraisal and suggestions from OSD executives, USAF officials, and SAB members. Then, from these many and varied responses, the SAB officers sought to readjust board structure and methods to the new times, yet keeping in mind Dr. Flax' injunction to take care that "what is already very good [in SAB] is not destroyed or disrupted."[13]

Of the many criticisms on hand when SAB officers began their discussions at a January 1963 Executive Committee meeting, three were perhaps most influential in charting the course of future action.[14] In one, Mr. John H. Rubel, Deputy Director

---

See notes on page 191.

of Defense Research and Engineering on Mr. McNamara's staff, explained why SAB space studies had failed to satisfy the needs of his office. In a presentation to the general board meeting in October 1962, Rubel criticized SAB space studies for not being sufficiently critical of Air Force planning in this area. Of the many SAB space studies he had read, he said, not one had suggested "that on balance the USAF ought to really give up Project X in order to do this more promising set of projects Y and Z." Since the Director of Defense Research and Engineering (DDR&E) had the responsibility to prepare a balanced program, he could not approve any recommendation "which merely generates desirements, which generates waste, which generates needs, all of which are additions for the largest program of its kind in the United States . . . [or which called for] vast undertakings in space that contribute little or nothing to our demonstrable military power, position or effectiveness."[15] Later, several top DDR&E officials afforded General Ferguson and Dr. Carter, the Air Force Chief Scientist, further insight into the criteria they applied for deciding the merits and feasibility of service proposals.[16]

A second critique came from Dr. Flax who observed "that approximately half of all SAB actions are treated on an ad hoc basis by interdisciplinary groups composed of representatives of more than one panel." He interpreted this as an indicator that "we tend to do better responding to problems submitted to us, rather than in generating problems of an ad hoc nature ourselves under our present form of organization." This was a weakness, he felt; the board ought to remain alert to upcoming problems of an interdisciplinary nature and generate its own studies of them. As a solution, he proposed

> that we create on the Executive Committee a series of assignments of individuals who would concern themselves with across-the-board system problems. These assignments would be superimposed on and parallel to present panel responsibilities of Executive Committee members. The purpose of these assignments would be to seek out the military systems problems which may require integrated, interdisciplinary across-the-board, and therefore, ad hoc action on the part of the SAB. This concept of SAB operation (or more exactly, a SAB posture to recognize these problems while operating in essentially the same manner as previously), should not be construed as a step in the direction of military systems analysis. Rather, the function of such a posture would be primarily to provide for the inter-disciplinary aspects of technology beyond the customary and "text book" (single discipline) look. It would provide a natural mechanism to anticipate problems

See notes on pages 191 and 192.

and generate new ideas other than those found solely within a specific area of technology.

Flax suggested, "for purposes of initiating discussion," the following as areas suited for assignment to individuals under his proposal: strategic systems, tactical systems (including limited war systems, counterinsurgency systems, and cold, political, and psychological warfare), aerospace defense systems, reconnaissance systems, and, finally, transport and logistic systems. Flax felt that his proposal would in no way change the present panel structure, that panels would still be "the most satisfactory way of marshalling technical personnel . . . [and for providing] the primary 'home' of the individual member."[17]

A third critique by Geophysics Panel chairman Paul A. Smith agreed that it was "timely to consider the structure and working methods of the SAB in considerably more depth than has been done for some years," and suggested that the first step should be "to get the answer to the question of whether, and if so why, the Air Force needs an advisory group now." In his own contemplations of the matter, he had attempted to balance SAB's assets to the Air Force against its liabilities. Included among the assets were: "(1) The opportunities [SAB provided] to scientists and military men for developing mutual confidence and respect; (2) assistance to Air Force officers in decisions involving a considerable element of judgment on applications of sci ntific principles to future Air Force operations; (3) wider aware. ss on the part of scientists of potential military applications a. 1 of military problems; and (4) support to the Air Force from ientists external to the Air Force organization on appropriate occasions." The cost of these (and other) SAB benefits to the Air Force included, in his opinion, "(1) the time and effort to keep the SAB informed of its activities and plans; (2) the time and effort to consider and apply them, or explain to the SAB the reasons why they could not do what the SAB recommended; (3) the timely selection of subjects appropriate to the SAB, and the arrangements for SAB consideration thereof; and (4) the complications and sometimes delays of finding SAB recommendations in conflict with or out of harmony with Air Force in-house analyses and decisions." If these assessments were correct, he asked, did they "make a demonstrably convincing case for continuing the SAB . . .?" Or, perhaps, was a different type of SAB now in order? In any event, whatever new directions the SAB took, it should

See notes on page 192.

strive to put "a higher premium on scientific and engineering judgment decisions properly balanced with military and political experience." In conclusion, Admiral Smith noted that[18]

> the most generally heard and valid criticisms have been the hasty and sometimes frenzied approaches to some pretty important matters dashed off by the SAB without spending enough time in reasoned consideration of the problems they are seized with. The best efforts I can recall have been those where the SAB groups have taken enough time to set down the pros and cons of alternative views, and given the varied views with supporting arguments to the Chief of Staff without trying to settle the impossible by vote, or without resorting to ambiguity for the sake of agreement.

Though they varied greatly in point and substance, the critiques had one significant point in common: SAB had to exercise more initiative to identify substantive tasks which it was singularly qualified to undertake. Toward this end, SAB officers spent 1963 in seeking out technical areas which promised "the most pay-off" (along lines suggested by Dr. Flax' critique), and in strengthening communication between SAB leadership and top Air Force personnel. As General Ferguson summarized the reason for and nature of this activity, "relatively few long-range conceptual problems [had been] put to the Board [in the past; now the Air Force desired] to get people thinking about a means for drawing the SAB into our long-range problems and surveys."[19]

By the end of the year, while the essential issue of "whither the board" was still far from resolved, SAB officers had taken several important steps concerning their studies. They had determined "to take on fewer jobs but do a better job on key problems." Further, on requests for studies, whether internally or externally proposed, they intended to identify the problem carefully, making certain it was appropriate for SAB, then stress quality rather than quantity in the finished product. In this connection, the Steering Committee resolved to winnow out excessive requests and make sure that the demands on SAB were specific.[20]

Secretary Zuckert and General LeMay allayed any doubts SAB officers may have had that the Air Force considered SAB's usefulness either diminished or at an end. At a meeting between them and the Steering Committee in January 1964, Dr. Stever first apprised them of the "feeling of dissatisfaction and frustration among the SAB members" that also seemed prevalent throughout the scientific community in recent months.

See notes on page 192.

Stever attributed this feeling to the current environment surrounding technology which seemed to have its roots in a number of ideas:

> (1) That we had reached a technology saturation point and that there was nothing else to do. (2) That there were advisory committees on everything these days and none of them seemed to have definite impact any more. (3) That possibly the scientific advisors were getting too old and regenerative in their ideas and that there was a need for young scientists. (4) That possibly scientific advisors were as good as ever but the Air Force and DOD now had an abundance of scientific talent of the SAB caliber such as RAND, Aerospace, MITRE, Lincoln, etc.

For his part, General LeMay answered, "he wanted to cheer up the board members and let them know that, in his opinion, their scientific advice has been good . . . that the board was doing all right and he was completely happy with its work. . . ." If its officers wished to change the board's size and procedures, they had his permission, but he suggested they institute change only if absolutely necessary. Mr. Zuckert agreed, adding that "in this period of limited or decreasing budgets, morale was bound to be lowered." However, he added very emphatically that "it is just this very time when the military services must retain a great nucleus of scientific strength to provide continuity . . . [and] maintaining the quality of the SAB is a necessity in order to keep this competence intact for the future."* Later, Dr. Flax, now serving as Assistant Secretary of the Air Force for Research and Development, noted his agreement with these statements. He could not "envision the time in the future when the Air Force [would] not have need for a SAB." As he viewed it:[21]

> The Secretary of the Air Force and the Chief of Staff have a continuous requirement for independent judgments on scientific and technical matters. The Secretary of the Air Force has a requirement to get diverse opinions, overviews, or a larger sphere of confidence on, not only problems that are specifically Air Force programs, but also problems that overlap other activities such as NASA, DOD, and the other services. The Secretary has a definite need for objective and informed judgment in the scientific and technical area. The SAB *must* maintain this capability. However, it must also be capable of analyzing and advising on programs solely within the Air Force although less and less programs have only Air Force areas of interest. We must crank in a larger degree of confidence in the SAB to tackle the specific problems.

---

*See quote at the beginning of this chapter.

See notes on page 192.

Following a discussion of SAB's efforts to isolate and concentrate on broad "substantive" problems, General LeMay proposed, and Mr. Zuckert agreed, that the SAB should set up working groups devoted to SAC, TAC, and ADC problems, with a fourth group keeping close check on anti-submarine warfare possibilities.[22]

By mid-January, the Steering Committee had translated this guidance into a proposed study approach which General Ferguson presented to the Executive Committee on the 21st. The proposal resulted from SAB's critical analysis throughout 1963 of its structure and functions "in the light of the current environment to determine ways of increasing its effectiveness to the Air Force," he said.   And it sought to implement SAB's conclusion that "the board could be most effective if it concentrated its future efforts on a few long-range studies of major importance to the Air Force . . . directed toward applying promising new technology in major mission areas of the Air Force."   The Steering Committee had considered the various subjects suited for such long-range, concentrated study and concluded that the following five seemed to offer the most promise: (1) tactical air, (2) missile base vulnerability, (3) anti-submarine warfare, (4) ballistic missile defense, and (5) military man-in-space. There were, however, several questions: should the SAB proceed in this way; which studies should it undertake first; and, perhaps most important, could it accomplish such broad studies within its current organization?[23]

The proposal generated a lengthy and far-ranging discussion among Executive Committee members. One member suggested that the DAG's could better undertake the proposed long-range studies—the BSD DAG handling missile vulnerability, the SSD DAG the man-in-space project, and the ASD and ESD DAG's the tactical air problem. Others felt that the DAG's could not do these and their regular jobs. There were also proposals for other study subjects.

On the question of organization, one member felt that "the present SAB panel structure is archaic in terms of today's Air Force needs . . . that the DAG's are directly analogous to the initial construction of the SAB," and suggested that the SAB reorganize along DAG lines. Another member supported the current structure viewing the panels as "holding" groups from which appropriate members can be selected to work on an ad hoc problem as required."[24]   In the end, the officers agreed to proceed with the new program and develop it within the current board structure.*

The Zuckert-LeMay and Executive Committee meetings of January 1964 marked the nadir of a two-year decline in SAB activity and spirit.   From this point the board moved forward with a steadily increasing ebullience of will and purpose.   The prime source of this renewed energy was the project bearing the title Tactical Air Capabilities Task Force, abbreviated to TAC Task Force.   As of late 1964, the new mode of operation manifested in this project seemed to have settled the question of "whither the board?" for some years to come.

The TAC Task Force project was born officially on January 28, 1964, when the SAB Steering Committee decided that of the five subjects considered for eventual study in depth, two—tactical air and military man-in-space—appeared most suited for first consideration.[25]   Soon after, Dr. Sheingold agreed to chair the tactical study and Dr. Gerald M. McDonnel the space study.   Initially, SAB officers thought to get the tactical study started then gradually phase in the space study —to proceed with both studies concurrently for a time, in other words.   As it turned out, however, once the TAC Task Force got under way, it became obvious that the SAB could handle only one study of such magnitude at a time.   Consequently, Dr. McDonnel's study remained on standby through the end of 1964.

To Dr. Sheingold fell the job of defining the goals of the TAC Task Force project, then of shaping an organization of suitable size and talent for attaining these goals.   By early March, he had decided that initially he would[26]

> establish a nucleus of key people . . . to look at the overall TAC area. We [will] then define special objectives which may be pertinent to the Air Force or the DOD [and] report these objectives to the [SAB] Steering Committee. . . . Based on its reaction to our initial efforts, we will then plan to establish a larger group of SAB and possibly outside people to look at the objectives which have been approved for us.

Soon after, Dr. Stever informed the SAB members that Dr. Sheingold was authorized to "call on . . . members of the board and Air Staff . . . to assist him in analyzing the major prob-

---

*In these early 1964 discussions of board structure, the Executive Committee initially elected to suspend the Arms Control Panel "until further requirements are established." Later that year, the committee dissolved it, thus reducing the number of panels to 10.

See notes on page 192.

lems in his area of study as a basis for defining [TAC Task Force] objectives and plan of action." This done, Sheingold would "proceed to establish a broader group, including special advisers . . . to insure proper coverage at all required areas of investigation."[27] Already it was assumed that the project would take six to twelve months, perhaps longer, to complete.

The TAC Task Force proceeded and progressed fairly closely to this initial concept. Beginning about mid-March 1964, Dr. Sheingold met with top OSD and Air Force officials to discuss the significant aspects of the subject. Meanwhile, the panel chairmen set out to determine how they and their colleagues might contribute.[28] On April 27-28, when the full board convened its spring 1964 meeting at Eglin Air Force Base, Florida, for, primarily, "a 'hardware' orientation" on tactical aviation, Sheingold briefed the panels on the project and discussed their ideas on it.

On May 15, General LeMay strongly endorsed "SAB's concentration on one of our most pressing problems—how to provide our Tactical Air Forces with the full technical capability to perform the tactical mission." In an accompanying tentative charter, he proposed that the group concentrate on exploiting "the technical competence of its members as a unique Air Force resource . . . [by emphasizing] the technical aspects of problems involving operational concepts." He then provided several examples of the aid which the task force could provide within this general guidance.[29] Later that month, Dr. Sheingold and members of the SAB Executive Committee discussed this tentative charter and organization possibilities with General Walter C. Sweeney, TAC commander and his staff and received a comprehensive briefing on TAC requirements. Then, at an Executive Committee meeting on May 31 in Colorado Springs, SAB officers decided on the organization. They prevailed on Mr. James F. Healey to serve as deputy chairman and created eight "working groups," named and chairmanned as follows" (1) Test and Evaluation—Mr. Horner; (2) Aircraft —Prof. Perkins; (3) Command and Control—Dr. Radford; (4) —Dr. Jack Ruina; (6) Weapons/Munitions—Dr. Latter; (7) Me-Navigation/Strike—Dr. Herwald; (5) Tactical Reconnaissance teorology—Dr. William W. Kellogg; (8) and Logistics—Dr. Murray A. Geisler. Later, they established a ninth working group

See notes on page 192.

—Engineering Geology under Dr. John C. Reed.* Finally, they created a task force Steering Group, which, as ultimately constituted, consisted of Dr. Sheingold, Mr. Healey, and the working group chairmen with the SAB Chairman, Dr. Stever, and the Air Force Chief Scientist, Dr. Winston P. Markey, as special advisors.[30]

By mid July, all working groups except Logistics were sufficiently manned to have held an initial organization and planning meeting with Dr. Sheingold. Manning the Logistics group lagged because of SAB's lack of experience in this area.[31] Consequently, as finally formed by late August, its total membership—including Chairman Geisler (RAND)—came from outside the SAB. Membership was mixed on the rest of the groups. By early fall 1964, task force membership totalled 80, of whom 37 were SAB members. Most of the others were from government agencies, universities, and such non-profit organizations as RAND and MITRE. A very few were in private industry. Additionally, except in a few instances where it was not feasible, the Air Staff, TAC, AFSC, OAR, and the Air Force Logistics Command assigned project officers to each group.[32] Some 40 officers (in ranks of major through colonel) were thus brought directly into TAC Task Force affairs, expanding total participation to about 120. General Schriever summarized the Air Force's satisfaction with this broad organization, noting his "extreme interest" in the approach on two accounts:[33]

> First, the wide participation which you have elicited will make it possible to generate greater understanding of the influence which technology will have on future operations. Second, the panel structure which you have organized should facilitate the identification of problem areas and avenues for further advances. During the course of your operation, I intend to insure that [AFSC] support is responsive to the importance of your work.

Mr. Zuckert shared this opinion, noting in a discussion with Dr. Sheingold that "he was pleased to see the military liaison members joining with the Air Staff in support of the work groups." This, he felt, "would surely assist in closing the gap between research and development and operational problems."[34] A final typical response came from Maj. Gen. Don R. Ostrander, OAR commander. OAR staff scientists attending the working group meetings could call on scientists in their

---

*This Working Group formed in August from a SAB ad hoc committee on military implications of engineering geology.

See notes on page 192.

own laboratories "so that OAR may couple more effectively with both the SAB and the needs of operating commands," he said.[35]

In late September, Dr. Sheingold issued a guidance document designed to serve, in effect, as "our charter of operation for the remainder of the effort." Since many groups earlier had explored the problems of tactical aviation and wrought improvement, the task force would not "repeat all the work already completed nor attack the entire problem in such detailed manner that sophisticated technical solutions are proposed which are out of context with the objective of achieving practical results." Rather, the goal was to "identify and suggest solutions to the most important technical problems affecting operations in a way that will result in timely decisions coupled with adequate financial support." In conclusion, Sheingold noted that the task force had no interest in defining roles and missions, that this was a job more properly belonging to military officers. Rather, the force's sole objective was to employ available technical backgrounds and experience toward helping the Air Force to solve "many critical problems which will lead to improvements in the capability of the Air Force to conduct tactical air operations in the future."[36]

Thus, on SAB's 20th birthday, more than one half of its members were direct participants, with the remainder on call, in what one officer described as "the first major effort orienting the SAB into a single major project."[37] The future course of the board could well be decided by the outcome of the TAC Task Force's final report. To do the job, Dr. Sheingold noted, "he had made every attempt to get good people who had knowledge and experience to work on the specific problems," and they were "looking for practical solutions, not necessarily the most complicated technological approach."[38] Target date for the finished report was still months away—not until July 1, 1965, at the earliest. Meanwhile, there had already been great benefits. As General Ferguson expressed it, "the fact that the SAB has interested itself in this area has shown all concerned that the Air Force is 'with' tactical air operations. The general climate has improved considerably so that timely, sound, and specific SAB inputs can be effective."[39]

In any event, the Air Force, in Mr. Zuckert's words, was "looking forward with great expectation" to the task force's report. At the same time, he observed that "this past year

See notes on page 192.

has been an enormously fruitful one in terms of SAB activity."
His concluding comment brought the SAB's first 20 years to
a graceful close: "Our gratitude to you all."[40]

See notes on page 192.

APPENDICES

A. Chart: SAB Executive and Administrative Officers

B. Roster: SAB Executive and Administrative Officers

C. SAB Membership Roster, Alphabetical

D. Changes in Panel Structure

E. SAB Membership Roster, by Assignment

F. SAB Steering and Executive Committees

G. SAB Membership on Division Advisory Groups

H. General Board Meetings (Dates and Locations)

I. Changes in Air Force Regulation 20-30

J. Studies and Authors, *Toward New Horizons*

K. Register of SAB Reports

EXECUTIVE AND ADMINISTRATIVE OFFICERS*
USAF SCIENTIFIC ADVISORY BOARD
1946-1964

Year	Executive Guidance			Executive Leadership †			Secretariat						SAB Membership
	Secretary of the Air Force	Assistant Sec of the Air Force (R&D)	Chief of Staff USAF	SAB Chairman	SAB Vice Chairman	SAB Military Dir.	Secretary	Technical Dir.	Asst Secy	Asst Secy	Asst Secy	Asst Secy	
1946			Spaatz				Johnson						30
1947	Symington						Alexander						32
1948						Craigie							34
1949			Vandenberg		Dryden		Walkowicz	Driscoll					35
1950				von Karman		Putt							46
1951	Finletter						Driscoll		Maier				51
1952					Doolittle Kelly	Craigie							61
1953							Hasert	Whitcraft	Summer				62
1954	Talbott	Gardner											59
1955			Twining	Kelly	Doolittle				Attinger				55
1956	Quarles			Doolittle		Putt	Sweet						55
1957		Horner							Gray				54
1958	Douglas				Stever		Duncan						51
1959		Charyk	White				Hasert	Hasert	Zubon				67
1960	Sharp	Perkins		Putt		Wilson	Gasser						75
1961		McMillan						Howard					77
1962	Zuckert		LeMay						Massey				88
1963				Stever	Perkins	Ferguson			Miller				68
1964		Flax					Burger		Hunt				68

* See Appendix B for full names.

† Excluding Panel Chairmen, Senior Statesmen, and Air Force Chief Scientists.
  See Appendices B, E, and F.

APPENDIX B

EXECUTIVE AND ADMINISTRATIVE OFFICERS*
USAF SCIENTIFIC ADVISORY BOARD
1946-1964

SECRETARY OF THE AIR FORCE
W. Stuart Symington	Sep 47-Apr 50
Thomas K. Finletter	Apr 50-Feb 53
Harold Talbott	Feb 53-Aug 55
Donald A. Quarles	Aug 55-Mar 57
James H. Douglas, Jr.	Mar 57-Dec 59
Dudley C. Sharp	Dec 59-Jan 61
Eugene M. Zuckert	Jan 61-

ASSISTANT SECRETARY OF THE AIR FORCE FOR RESEARCH AND DEVELOPMENT
Trevor Gardner	Mar 53-Feb 56
Richard E. Horner	Mar 56-May 59
Joseph V. Charyk	Jun 59-Jan 60
Courtland D. Perkins	Apr 60-Jan 61
Brockway McMillan	Jan 61-Jun 63
Alexander H. Flax	Jul 63-

CHIEF OF STAFF, UNITED STATES AIR FORCE
General Carl A. Spaatz	Sep 47-Apr 48
General Hoyt S. Vandenberg	Apr 48-Jun 53
General Nathan F. Twining	Jun 53-Jun 57
General Thomas D. White	Jul 57-Jun 61
General Curtis E. LeMay	Jun 61-

CHIEF SCIENTIST, UNITED STATES AIR FORCE
Louis N. Ridenour	Sep 50-Aug 51
David T. Griggs	Dec 52-Jun 53
Chalmers W. Sherwin	Feb 54-Jan 55
H. Guyford Stever	Feb 55-Jul 56
Courtland D. Perkins	Aug 56-Jul 57
George E. Valley, Jr.	Sep 57-Dec 58
Joseph V. Charyk	Jan 59-Jun 59
Alexander H. Flax	Oct 59-Apr 61
Leonard S. Sheingold	Jul 61-Jul 62
Launor F. Carter	Jul 62-Jul 63
Robert W. Buchheim	Jul 63-Jun 64
Winston R. Markey	Aug 64-

CHAIRMAN, USAF SCIENTIFIC ADVISORY BOARD
Theodore von Karman	Mar 46-Dec 54
Mervin J. Kelly	Jan 55-Nov 55
James H. Doolittle	Nov 55-Dec 58
Donald L. Putt	Jan 59-Dec 61
H. Guyford Stever	Jan 62-

VICE CHAIRMAN, USAF SCIENTIFIC ADVISORY BOARD
Hugh L. Dryden	Apr 48-Dec 50

*Excluding panel chairmen and Senior Statesmen (see Appendix E).

Mervin J. Kelly	Jan 51-Dec 54
James H. Doolittle	Jan 51-Nov 55
H. Guyford Stever	Aug 56-Dec 61
Courtland D. Perkins	Jan 62-

MILITARY DIRECTOR, USAF SCIENTIFIC ADVISORY BOARD

Maj Gen Laurence C. Craigie	Apr 48-Sep 48
Maj Gen Donald L. Putt	Sep 48-Jan 52
Lt Gen Craigie	Jan 52-Apr 54
Lt Gen Putt	Apr 54-Jul 58
Lt Gen Roscoe C. Wilson	Jul 58-Nov 61
Lt Gen James Ferguson	Nov 61-

COMMANDER, ARDC/AFSC*

Maj Gen David M. Schlatter	Feb 50-Jun 51
Lt Gen Earle E. Partridge	Jun 51-Jun 53
Lt Gen Donald L. Putt	Jun 53-Apr 54
General Thomas S. Power	Apr 54-Jun 57
Maj Gen John W. Sessums, Jr.	Jul 57-Jul 57
Lt Gen Samuel E. Anderson	Aug 57-Mar 59
Maj Gen John W. Sessums, Jr.	Mar 59-Apr 59
General Bernard A. Schriever	Apr 59-

SECRETARIAT OFFICERS, USAF SCIENTIFIC ADVISORY BOARD

Secretary

Ralph P. Johnson (additional duty)	Jul 46-Feb 47
Maj Donald M. Alexander	Feb 47-Nov 48
Lt Col Teddy F. Walkowicz	Nov 48-Nov 50
Mr. B. J. Driscoll	Nov 50-Nov 52
Mr. Chester N. Hasert	Nov 52-Feb 55
Lt Col Floyd J. Sweet	Feb 55-Sep 57
Col George H. Duncan	Sep 57-Nov 58
Col Clyde D. Gasser	Nov 58-Jun 63
Col Robert J. Burger	Jun 63-

Technical Director

Mr. Chester N. Hasert	Feb 55-

Assistant Secretaries

Mr. B. J. Driscoll	Jun 49-Nov 50
Maj Mark P. Maier	Apr 50-Jul 53
Maj Daniel D. Whitcraft, Jr.	Jun 52-Jul 55
Capt James A. Summer	Jul 53-Jan 54
Capt Frank S. Attinger	Jan 54-Oct 57
Lt Col Billy C. Gray	Jul 55-Aug 59
Lt Col Michael Zubon	Oct 57-Aug 61
Lt Col Kent P. Howard	Aug 59-
Lt Col Julius H. Massey, Jr.	Sep 60-
Col James E. Miller	Aug 61-
Lt Col Senour Hunt	Aug 64-

Administrative Assistants

Mrs. Marie Roddenberry	1945-Oct 51
Mrs. June Merker	Oct 51-Oct 52
Mrs. Adelia Letchworth	Oct 52-

---

*Served as ex-officio board members from July 1956 to November 1962. From that date served as members of the SAB Executive Committee (see Appendix F).

NAME	46	47	48	49	50	51	52	53	54	55	56	57	58	59	60	61	62	63	64	
Adey, Dr. W. Ross																	■			UCLA Medical Center
Agnew, Dr. Harold M.															■	■	■	■		Los Alamos Scientific Lab.
Ashley, Dr. Holt																		■		MIT
Astin, Dr. Allen V.						■	■	■	⊠	⊠	⊠	⊠	⊠							Nat'l. Bureau of Standards
Baily, Dr. Norman A.														■	■	■	■			Hughes Research Labs.
Baker, Dr. James G.					■	■	■	■	■	■										Harvard Univ.
Baldes, Dr. Edward J.																				Mayo Clinic
Barlow, Mr. Edward J.					■															Aerospace Corp.
Bartky, Dr. Walter					■	■														Univ. of Chicago
Battan, Prof. Louis J.																■	■			Univ. of Ariz.
Berlin, Mr. Don R.																				McDonnel Acft Corp.
Bethe, Prof. Hans A.																		■		Cornell Univ.
Billings, Dr. Bruce																				Aerospace Corp.
Bisplinghoff, Prof. Raymond L.								■	■	■	■	■	■							MIT
Bogdonoff, Prof. Seymour																				Princeton Univ.
Boilay, Dr. William					■	■		■	■	■										Aerophysics Dev. Corp.
Bondurant, Dr. Stuart																		■		Indiana Univ. Medical Center
Bradbury, Dr. Norris E.																				Los Alamos
Bray, Dr. Charles W.																				Princeton Univ.
Bronk Dr. Detlev W.		■	■	■	■	■	■	■	⊠	⊠	⊠	⊠								Nat'l Academy of Sci.
Brown, Dr. J. Douglas																				Princeton Univ.
Brown, Dr. Harold													■	■						Lawrence Rad. Lab.
Brueckner, Prof. Keith A.												■	■	■				■		Inst. for Defense Analysis
Buchon, Dr. Roland F.																				Prudential Ins. Co.
Burchell, Dr. Howard B.																				Mayo Clinic
Carlson, Dr. Loren D.									■	■	■	■	■	■	■					Univ. of Ky. School of Medicine
Carter, Dr. Launor F.										■	■		■	■	■	■				Systems Dev. Corp.
Charyk, Dr. Joseph V.							■	■	■	■	■									Aeronutronic Systems, Inc.
Clauser, Dr. Francis H.													■							Johns Hopkins Hosp.
Clauser, Dr. Milton U.											■	■	■	■						Space Technology Labs.
Cottrel, Dr. Leonard S.																				Russel Sage Foundation
Cook, Dr. Thomas B.																		■		Sandia Corp.

* ■ Years of membership. ⊠ Ex officio member or associate advisor.

136

APPENDIX C (Continued)

NAME	46	47	48	49	50	51	52	53	54	55	56	57	58	59	60	61	62	63	64	
Dinneen, Dr. Gerald P.																				MIT
Donovan, Dr. Allen F.																				Aerospace Corp.
Doolittle, Dr. James H.																				Space Technology Labs.
Dollard, Mr. Charles																				Carnegie Corp.
Draper, Dr. Charles S.																				MIT
Dryden, Dr. Hugh L.																				NASA
Dubridge, Dr. Lee A.																				Caltech
Dunn, Dr. Louis G.																				Caltech
Duwez, Dr. Pol E.																				Caltech
Eggers, Dr. Alfred J., Jr.																				NASA
Ellett, Dr. Alexander																				Zenith Radio Corp.
Elvey, Dr. C(hristian) T.																				Univ. of Alaska
Emmons, Dr. Howard W.																				Harvard Univ.
Fermi, Prof. Enrico																				Univ. of Chicago
Ferri, Dr. Antonio																				New York Univ.
Fitts, Dr. Paul M.																				Univ. of Mich.
Flax, Dr. Alexander H.																				Cornell Aero. Lab.
Fletcher, Dr. J(ames) C.																				Univ. of Utah
Flickinger, B/G (Ret) Don																				Private Consultant
Foster, Dr. John S., Jr.																				Lawrence Rad. Lab.
Friis, Dr. Harold T.																				Bell Telephone Co.
Frische, Dr. Carl A.																				Sperry Gyro. Co.
Fubini, Dr. Eugene																				Airborne Inst. Lab.
Fultz, Dr. Dave																				Univ. of Chicago
Gamow, Dr. George A.																				Johns Hopkins Univ.
Gardner, Dr. John W.																				Carnegie Corp.
Getting, Dr. Ivan A.																				Aerospace Corp.
Gibson, Dr. Ralph E.																				Applied Physics Lab.
Gilruth, Dr. Robert R.																				NACA
Goldberg, Prof. Leo																				Harvard Univ.
Gray, Mr. William L.																				Boeing Co.
Griggs, Prof. David T.																				Univ. of Calif. at L.A.

APPENDIX C (Continued)

NAME	46	47	48	49	50	51	52	53	54	55	56	57	58	59	60	61	62	63	64	
Gunn, Dr. Ross																				U. S. Weather Bureau
Harris, Dr. Payne S.																				Private Consultant
Hartig, Dr. Elmer																				Goodyear Aerospace Corp.
Hastings, Dr. Donald W.																				Univ. of Minn.
Haurwitz, Prof. Bernhard																				New York Univ.
Hawthorne, Prof. William R.																				MIT
Healey, Mr. James F.																				Honeywell
Herbst, Dr. Roland F.																				Lawrence Rad. Lab.
Herwald, Dr. S(eymour) W.																				Westinghouse Elec.- Corp.
Hickam, Dr. John B.																				Duke Univ.
Hoff, Prof. Nicholas J.																				Stanford Univ.
Holaday, Mr. William M.																				Socony Vacuum Oil Co.
Holloway, Dr. Marshall G.																				ACF Industs. Inc.
Holzer, Prof. Robert E.																				UCLA
Holzman, Col. Benjamin G.																				USAF
Horner, Mr. Richard E.																				Northrop Space Labs.
Houghton, Dr. Henry G.																				USAF
Hovland, Dr. Carl I.																				Yale Univ.
Huffman, Prof. David A.																				MIT
Humphreys, Dr. Lloyd G.																				Univ. of Ill.
Hunter, Dr. Walter S.																				Brown Univ.
Hutcheson, Dr. John A.																				Westinghouse Res. Lab.
Hyatt, Mr. Abraham																X				NASA
Iklé, Dr. Fred C.																				RAND
Jones, Dr. Loren F.																				RCA
Kalitinsky, Dr. Andrew																				M. W. Kellogg Co.
Kaplan, Prof. Joseph																				UCLA
Kaufmann, Prof. William W.																				MIT
Kellogg, Dr. William W.																				Nat'l. Center for Atmospheric Res.
Kelly, Dr. Mervin J.																				Bell Telephone Co.
Kent, Dr. Robert H.																				Aberdeen Proving Ground
King, Dr. Gilbert W.																				ITEK Corp.

APPENDIX C (Continued)

NAME	46	47	48	49	50	51	52	53	54	55	56	57	58	59	60	61	62	63	64	
Kistiakowsky, Dr. George B.															■					Harvard Univ.
Kossman, Dr. Charles E.																				New York Univ.
Land, Dr. Edwin H.							■	■	■	■										Polaroid Corp.
Langmuir, Dr. Irving												■								Gen'l. Electric Co.
Lanphier, Mr. Thomas G., Jr.				■	■															Gen'l. Electric Co.
Latter, Dr. Albert L.											■				■					RAND
Lauritsen, Dr. Charles C.										■	■									Caltech
Lawson, Dr. James L.														■						Gen'l. Electric Co.
Leghorn, Dr. Richard C.							■						■							ITEK Corp.
Licklider, Dr. J(oseph) C. R.																	×			ARPA
Lindsley, Dr. Donald B.															■					Northwestern Univ.
Lion, Dr. Kurt S.																				MIT
Long, Dr. F(ranklin) A.											■			■	■					Cornell Univ.
Longmire, Dr. Conrad									■											Los Alamos Sci. Lab.
Love, Mr. Mark P. L., Jr.										■										Shell Oil Co.
Lovelace, Dr. W. Randolph, II													■	■						Lovelace Foundation
Macdonald, Dr. Duncan E.									■											ITEK Corp.
MacDougall, Dr. Duncan P.			■																	Naval Ordn. Lab.
Macelwane, Dr. James B.																				St. Louis Univ.
MacNair, Dr. Walter A.														■						Bell Telephone Co.
Madden, Mr. John D.															■					Assn. for Computing Mach.
Malone, Dr. Thomas F.														■	■					Travelers Ins. Co.
Marbarger, Dr. John P.															■					Univ. of Ill.
Mark, Dr. J. Carson						■	■	■												Los Alamos
Markham, Dr. John R.																				MIT
McCormack, M/G James, USAF. (Ret)													×		■	■				MIT
McDonnel, Dr. Gerald M.																■				UCLA Medical Center
McMillan, Dr. William G.											■				■	■				RAND
Meyer, Dr. James W.																■				MIT
Millar, Mr. Julian Z.																				Western Union
Miller, Dr. Burton F.																				Thompson–Ramo–Wooldridge, Inc.
Miller, Dr. Carlton W.												■								Perkin–Elmer Corp.

APPENDIX C (Continued)

NAME	46	47	48	49	50	51	52	53	54	55	56	57	58	59	60	61	62	63	64	
Miller, Dr. George A.																				Harvard Univ.
Miller, Prof. Rene H.																				MIT
Miller, Mr. Stewart E.																				Bell Telephone Labs.
Millikan, Dr. Clark B.																				Caltech
Millikan, Prof. Max F.																				MIT
Mills, Dr. Mark M.																				Univ. of Calif.
Molnar, Dr. J(ulius) P.																				Bell Telephone Co.
Moore, Mr. John R.																				No. Am. Aviation
Morgan, Dr. Clifford T.																				Private Consultant
Morton, Dr. George A.																				David Sarnoff Res. Center
Newmark, Dr. Nathan																				Univ. of Ill.
Norton, Prof. Dee W.																				State Univ. of Iowa
O'Brien, Dr. Brian																				Consulting Physicist
Oliver, Dr. Jack E.																				Lamont Geol. Obs.
Orlansky, Dr. Jesse																				Inst. for Defense Analysis
Overhage, Dr. Carl F. J.																				Eastman Kodak Co.
Partridge, Gen. Earle E., USAF (Ret)																				Lockheed Acft. Corp.
Peter, Mr. Marc, Jr.																				E. H. Plesset, Assoc.
Peterson, Dr. Lysle H.																				Bockus Res. Inst.
Perkins, Prof. Courtland D.																				Princeton Univ.
Pickering, Dr. William H.																				Caltech
Pigford, Dr. Robert L.																				Univ. of Delaware
Plesset, Dr. Ernst H.																				E. H. Plesset, Assoc.
Plesset, Dr. Milton S.																				Caltech
Pollard, Dr. Ernest C.																				Yale Univ.
Pool, Dr. Ithiel de Sola																				MIT
Porter, Dr. Richard W.																				Gen'l. Electric Co.
Pratt, Mr. Perry W.																				United Acft. Corp.
Princi, Dr. Frank																				Univ. of Cincinnati School of Medicine
Purcell, Dr. Edward M.																				Harvard Univ.
Putt, L/G Donald L. USAF (Ret)																				United Acft. Corp.
Quarles, Dr. Donald A.																				Sandia Corp.

APPENDIX C (Continued)

NAME	46	47	48	49	50	51	52	53	54	55	56	57	58	59	60	61	62	63	64	
Radford, Dr. William H.																				MIT
Rambo, Dr. William R.																				Stanford Univ.
Ramo, Dr. Simon																				Ramo-Wooldridge, Inc.
Ramsey, Dr. Norman F.																				Harvard Univ.
Rannie, Prof. William D.																				Caltech
Raymond, Dr. Richard C.																				Gen'l. Electric Co.
Reed, Dr. John C.																				Arctic Inst. of No. Am.
Ridenour, Dr. Louis N.																				Univ. of Ill.
Robertson, Dr. Howard P.																				Caltech
Root, Dr. L. Eugene																				Lockheed Acft. Co.
Rosenblith, Prof. Walter A.																				MIT
Rothrock, Mr. Addison M.																				NACA
Ruina, Dr. Jack																				Institute for Defense Analysis
Scheirer, Mr. George S.																				Boeing Acft. Co.
Schelling, Prof. Thomas C.																				Harvard Univ.
Schmued, Mr. Edgar																				Private Consultant
Schilling, Dr. Gerhard																				RAND
Schrieber, Dr. R (oener) E.																				Los Alamos
Scoville, Dr. Herbert S., Jr.																				U. S. Gov't.
Seamans, Dr. Robert C., Jr.																			X	NASA
Sears, Dr. William R.																			X	Cornell Univ.
Shank, Mr. Robert U.																				FAA
Shanley, Prof. Francis R.																				UCLA
Shapiro, Prof. Ascher H.																				MIT
Sheingold, Dr. Leonard S.																				Sylvania Elec. Sys.
Sherwin, Dr. Chalmers W.																				Aerospace Corp.
Shockley, Dr. William								X												Shockley Transistor Corp.
Siegel, Prof. Keeve M.																				Conductron Corp.
Silverstein, Dr. Abraham																				NASA
Singleton, Dr. James W.																				Systems Dev. Corp.
Smith, Mr. C. Branson																				United Acft. Corp.
Smith, R/Adm. Paul A. C.&GS (Ret)																				RAND

APPENDIX C (Continued)

NAME	46	47	48	49	50	51	52	53	54	55	56	57	58	59	60	61	62	63	64	
Soderberg, Prof. C. Richard																				MIT
Speir, Dr. Hans																				RAND
Spilhaus, Dr. Athelstan F.																				Univ. of Minn.
Stalnaker, Dr. John M.																				Assn. of Am. Med. Col.
Starr, Dr. Chauncey																				No. Am. Aviation Co.
Stevens, Mr. Frederick																				Northrop Nortronics
Stever, Dr. H. Guyford									⊠											MIT
Stewart, Dr. Homer J.																				Caltech
Stratton, Prof. Julius A.																				MIT
Street, Prof. J(abez) C.																				Harvard Univ.
Strong, Dr. John D.																				Johns Hopkins Univ.
Strong, Mr. Philip G.																				U. S. Gov't.
Sutton, Mr. George P.																				Rocketdyne
Swanson, Mr. Warren E.																				No. Am. Aviation, Inc.
Sweeney, Dr. William J.																				Standard Oil Co.
Taylor, Prof. Edward S.																				MIT
Taylor, Mr. Phillip B.																				Sanderson & Porter
Teller, Dr. Edward																				Lawrence Rad. Lab.
Thompson, Dr. Louis T. E.																				Norden Lab. Corp.
Thompson, Mr. Thomas H.																				Bellcom, Inc.
Thom, Dr. Robert N.																				Los Alamos Sci. Lab.
Tinus, Dr. William C.																				Bell Telephone Lab.
Tonge, Dr. Frederic M.																				Univ. of Calif. at Irvine
Townes, Dr. Charles H.																				MIT
Tsien, Dr. Hsue-shen																				Caltech
Tullis, Dr. John L.																				New Jersey Hosp.
Valley, Dr. George E., Jr.														⊠						MIT
Villard, Dr. Oswald G.																				Stanford Univ.
von Karman, Dr. Theodore H.																				AGARD
von Neumann, Dr. John																				Princeton Univ.
Wagner, Dr. Bernard M.																				New York Medical College

APPENDIX C (Continued)

NAME	46	47	48	49	50	51	52	53	54	55	56	57	58	59	60	61	62	63	64	
Walkowicz, Mr. Teddy F.																				Rockefeller Foundation
Ware, Dr. Willis H.																				RAND
Warren, Dr. Shields																				
Watson, Prof. Kenneth M.																				Boston Hospital
Wattendorf, Dr. Frank L.																				UCLA
Weinberg, Dr. Alvin M.																				AGARD
Weiss, Mr. Herbert K.																				Oak Ridge
Wexler, Dr. Harry																				Aerospace Corp.
Wheelon, Dr. Albert D.																				U. S. Weather Bureau
Wheeler, Prof. John A.																				U. S. Gov't.
Whipple, Dr. Fred L.																				Princeton Univ.
White, Dr. Clayton S.																				Smith. Astro. Obs.
Williams, Prof. Robin M.																				Lovelace Foundation
Wolfle, Dr. Dael																				Cornell Univ.
Wrigley, Dr. Walter W.																				Am. Assn. for Adv. Sci.
York, Dr. Herbert F.																				MIT
Young, Mr. Gale																				Lawrence Rad. Lab.
Young, Dr. Lloyd A.																				Nuclear Dev. Assoc.
Zwicky, Dr. Fritz																				RAND
Zworykin, Dr. Vladimir K.																				Caltech
																				RCA

CHANGES IN PANEL STRUCTURE *

	Fiscal Year						Calendar Year											
	47	48	49	50	51	52	53	54	55	56	57	58	59	60	61	62	63	64
Aeromedicine and Psychology																		
Aeromedicine and Social Sciences																		
Aeromedicine																		
Aeromedical Research																		
AEROMEDICAL/BIOSCIENCES																		
Aircraft and Propulsion																		
Aircraft - Fuels and Propulsion																		
Aircraft																		
AEROSPACE VEHICLES																		
ARMS CONTROL																		
BASIC RESEARCH																		
Radar, Communications, Weather																		
Electronics and Communications																		
ELECTRONICS																		
Weather and Upper Air Research																		
Geophysical Research																		
GEOPHYSICS																		
Fuels, Explosives and Nuclear Energy																		
Explosives and Nuclear Energy																		
Explosives and Armament																		
GUIDANCE AND CONTROL																		
Guided Missiles - Pilotless Aircraft																		
Guided Missiles																		
INFORMATION PROCESSING																		
Nuclear Weapons																		
NUCLEAR																		
Fuels and Propulsion																		
PROPULSION																		
Physical Sciences																		
Intelligence Systems																		
Reconnaissance																		
Social Sciences																		
PSYCHOLOGY & SOCIAL SCIENCES																		
Space Technology																		
Number of Panels	5	6	6	8	9	9	9	9	9	9	9	9	11	11	12	11	11	10
Number of Members on Board	32	34	35	46	51	61	62	59	55	55	54	51	67	75	77	88	68	68

* Large print denotes panels operational in 1964; small print denotes
those discontinued. Shaded areas depict the years panels were operational.

APPENDIX E
SAB MEMBERSHIP ROSTER: BY ASSIGNMENT*

SENIOR STATESMEN

Name	46	47	48	49	50	51	52	53	54	55	56	57	58	59	60	61	62	63	64
Charles S. Draper														■	■	■	■	■	■
Ivan A. Getting														■					
Joseph Kaplan														■	■	■	■	■	■
Clark B. Millikan														■	■	■	■	■	■
Edward Teller														■	■	■	■	■	■
Shields Warren														■	■		■		
Frank Wattendorf														■	■	■	■	⊞	

AEROMEDICAL/BIOSCIENCES PANEL

Name	46	47	48	49	50	51	52	53	54	55	56	57	58	59	60	61	62	63	64
W. Ross Adey																		■	■
Norman A. Baily																		■	■
Edward J. Baldes						■	■	■	■	■									
Stuart Bondurant																		■	■
Charles W. Bray						■	■	■									■	■	■
Detlev W. Bronk					■	■	■												
Ronald F. Buchan																	■	■	
Howard B. Burchell													■	■	■	■			
Loren D. Carlson													■	■	■	⊠	⊠	⊠	
Paul M. Fitts									■	■	■	■	★						
Don Flickinger																	■	■	
Payne S. Harris													■	■	■	■			
Donald W. Hastings						■	■	⊠	⊠	⊠	■								
John B. Hickam						■	■	■	■	⊠	⊠								
Walter S. Hunter					■	■	■	■	■	■									
Charles E. Kossman											■	■	■	■					
Donald B. Lindsley				■	■														
Kurt S. Lion																■	■		
W. Randolph Lovelace, II	⊠	⊠	⊠	⊠	⊠			■											
John P. Marbarger														■	■	■			
Gerald M. McDonnel														■	■			⊠	⊠
Clifford T. Morgan															★				
Lysle H. Peterson														■	■	■			
Frank Princi														■	■	■			
John L. Tullis										■	■	■	■						
Bernard M. Wagner													■	■					
Shields Warren						■	■	■	■	■									
Clayton S. White												⊠	⊠	⊠					

AEROSPACE VEHICLES PANEL

Name	46	47	48	49	50	51	52	53	54	55	56	57	58	59	60	61	62	63	64
Holt Ashley													■			■	■	■	■
Allen V. Astin								■											
Edward J. Barlow												■	⊠						
Don R. Berlin					■	■													
Raymond L. Bisplinghoff										■	■	■	■	■					
Seymour Bogdonoff														■	■	■	■	■	■
William Bollay				■	■	■													
Allen F. Donovan						■	■	■	■	■	■								
Pol E. Duwez			■	■	■														

* ■ - Years of membership.   ⊠ - Tenure as chairman.   ★ - Liaison member.

(See Appendix D for precise panel names for particular years.)

APPENDIX E (Continued)

AEROSPACE VEHICLES PANEL
(Continued)

	46	47	48	49	50	51	52	53	54	55	56	57	58	59	60	61	62	63	64
Francis H. Clauser						■	■	■	■	■	■	■							
Milton U. Clauser															■				
Alfred J. Eggers, Jr.													■	■					
Robert R. Gilruth													■					■	
William L. Gray																		■	■
William R. Hawthorne			■	■														■	
Nicholas J. Hoff																		■	
Richard E. Horner																		■	
John R. Markham								■											
Rene H. Miller																■	■	■	
Clark B. Millikan								⊠	⊠	⊠	⊠								
Courtland D. Perkins		■												⊠	⊠		⊠	⊠	⊠
Perry W. Pratt														★	★	★			
L. Eugene Root						⊠	⊠		★	★	■								
George S. Schairer																			
William R. Sears		■			⊠														
Francis R. Shanley							■	■											
C. Richard Soderberg	⊠	⊠	⊠																
Homer J. Stewart									■										
Warren E. Swanson																	■		
William J. Sweeney																			
Hsue-shen Tsien	■	■	■																
Frank L. Wattendorf																			

ARMS CONTROL PANEL †

	46	47	48	49	50	51	52	53	54	55	56	57	58	59	60	61	62	63	64
Launor F. Carter															⊠				
John S. Foster, Jr.															■				
Fred C. Ikle															■				
Albert L. Latter															■				
Richard C. Leghorn															■				
James McCormack															■				
Carl F. J. Overhage															■				
Thomas C. Schelling																⊠			
Herbert S. Scoville, Jr.															■				
Charles H. Townes															■				
Teddy F. Walkowicz															■				

BASIC RESEARCH PANEL

	46	47	48	49	50	51	52	53	54	55	56	57	58	59	60	61	62	63	64
Bruce Billings																			
Alexander H. Flax																★	★		
George Goldberg															★	★	★	★	
J.C.R. Licklider															■				
Brian O'Brien																		■	
William Shockley																★	★	★	
Charles H. Townes															★	★			
George E. Valley, Jr.															⊠	⊠	⊠	⊠	⊠

† Designated Arms Control Committee in 1961-1962.

APPENDIX E (Continued)

ELECTRONICS PANEL

	46	47	48	49	50	51	52	53	54	55	56	57	58	59	60	61	62	63	64
Edward J. Barlow														★	★				
Gerald P. Dinneen																	★		
Lee A. Dubridge																			
Alexander Ellett																			
J. C. Fletcher																			
Harold T. Friis																			
Eugene G. Fubini																			
Ivan A. Getting																			
Marshall G. Holloway																			
Benjamin G. Holzman																			
David A. Huffman																			
Loren F. Jones																			
Mervin J. Kelly																			
Gilbert W. King																			
James L. Lawson																			
James W. Meyer																			
Julian Z. Millar																			
Burton F. Miller																			
George A. Morton																			
Ernest C. Pollard																			
Richard W. Porter																			
Edward M. Purcell																			
Donald A. Quarles																			
William H. Radford																			
Simon Ramo																			
Louis N. Ridenour																			
Jack Ruina																			
Leonard S. Sheingold																			
Chalmers W. Sherwin																			
William Shockley																			
Keeve M. Siegel														★	★	★			
Julius A. Stratton																			
J. C. Street																			
William C. Tinus																			
Charles H. Townes																			
George E. Valley, Jr.														★	★	★			
Oswald G. Villard																		★	
Lloyd A. Young																			
Vladimir K. Zworykin																			

GEOPHYSICS PANEL

Louis Battan																			
C. T. Elvey																			
David Fultz																			
Leo Goldberg																			
Ross Gunn																			
Bernhard Haurwitz																			
Robert E. Holzer																			
Benjamin G. Holzman																			

APPENDIX E (Continued)

GEOPHYSICS PANEL (Continued)

	46	47	48	49	50	51	52	53	54	55	56	57	58	59	60	61	62	63	64
Henry G. Houghton																			
Joseph Kaplan																			
William W. Kellogg																			
Irving Langmuir																			
James B. Macelwane																			
Thomas F. Malone																			
Dee W. Norton																			
Jack E. Oliver																			
Richard W. Porter																			
John C. Reed																			
Gerhard Schilling																			
Keeve M. Siegel																			
Paul A. Smith																			
Athelstan F. Spilhaus																			
John D. Strong																			
Oswald G. Villard																			
Harry Wexler																			
Albert D. Wheelon																			
Fred L. Whipple																			
Walter W. Wrigley																			
Vladimir K. Zworykin																			

GUIDANCE AND CONTROL PANEL

	46	47	48	49	50	51	52	53	54	55	56	57	58	59	60	61	62	63	64
Walter Bartky																			
Norris Bradbury																			
Detlev W. Bronk																			
Charles S. Draper																			
Lee A. Dubridge																			
Enrico Fermi																			
Carl A. Frische																			
George A. Gamow																			
Elmer Hartig																			
James F. Healey																			
S. W. Herwald																			
John A. Hutcheson																			
Robert H. Kent																			
Charles C. Lauritsen																			
F. A. Long																			
Duncan P. MacDougall																			
Carlton W. Miller																			
John R. Moore																			
Nathan Newmark																			
Dee W. Norton																			
Marc Peter, Jr.																			
Edward M. Purcell																			
Norman F. Ramsey																			
Howard P. Robertson																			
Edgar Schmued																			
Robert C. Seamans, Jr.																			

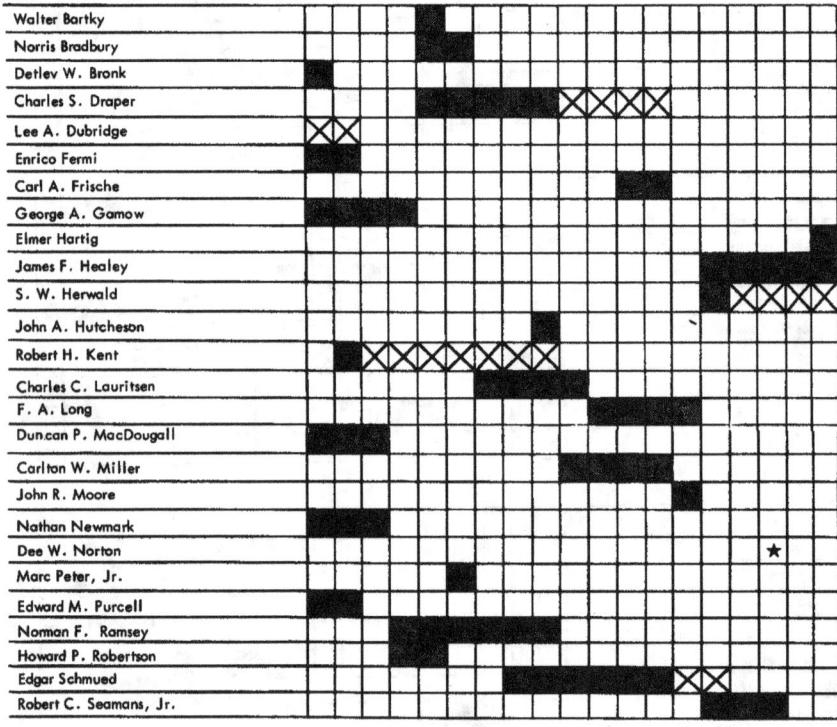

APPENDIX E (Continued)
GUIDANCE AND CONTROL PANEL (Continued)

	46	47	48	49	50	51	52	53	54	55	56	57	58	59	60	61	62	63	64
Robert J. Shank																			
Leonard S. Sheingold															★				
Keeve M. Siegel																			
Frederick Stevens																			
H. Guyford Stever																			
William J. Sweeney																			
Louis T. E. Thompson																			
Thomas H. Thompson																			
Herbert K. Weiss																			
Walter W. Wrigley																			
Fritz Zwicky																			

GUIDED MISSILES PANEL

Allen V. Astin																			
William Bollay																			
Charles S. Draper																			
Hugh L. Dryden																			
Ivan A. Getting																			
Ralph E. Gibson																			
Robert R. Gilruth																			
Robert H. Kent																			
Walter A. MacNair																			
John R. Markham																			
William H. Pickering																			
H. Guyford Stever																			
Homer J. Stewart																			

INFORMATION PROCESSING PANEL

Gerald P. Dinneen																			
Gilbert W. King																			
John D. Madden																			
Burton F. Miller																			
Earle E. Partridge																★			
Walter A. Rosenblith																			
Frederic M. Tonge																			
Willis H. Ware																			

NUCLEAR PANEL

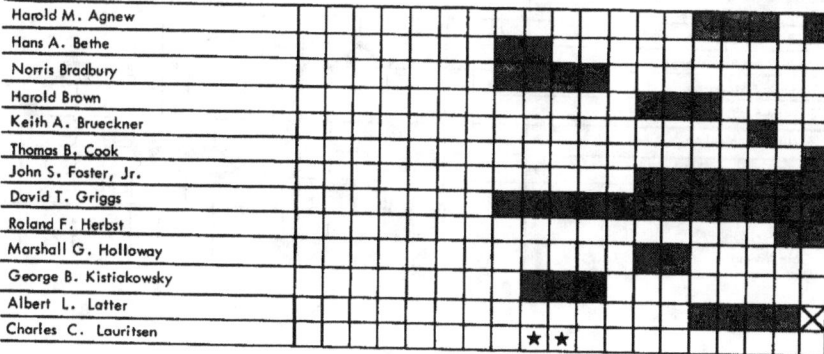

Harold M. Agnew																			
Hans A. Bethe																			
Norris Bradbury																			
Harold Brown																			
Keith A. Brueckner																			
Thomas B. Cook																			
John S. Foster, Jr.																			
David T. Griggs																			
Roland F. Herbst																			
Marshall G. Holloway																			
George B. Kistiakowsky																			
Albert L. Latter																			
Charles C. Lauritsen									★	★										

APPENDIX E (Continued)

NUCLEAR PANEL (Continued)

	46	47	48	49	50	51	52	53	54	55	56	57	58	59	60	61	62	63	64
Conrad Longmire																		■	■
J. Carson Mark																■	■	■	■
William G. McMillan																	■	■	■
Mark M. Mills											★	★							
J. P. Molnar														■	■				
Ernst H. Plesset												■	✕	✕	✕	✕	✕	✕	
L. Eugene Root								■	■	■	■	■							
Herbert S. Scoville, Jr.												★	■	■					
Chauncey Starr								■	■							■			
Edward Teller								■	■	■	✕	✕							
Robert Thorn																			
John von Neumann							✕	✕	✕										
Kenneth Watson																■	■	■	■
Alvin Weinberg													■	■	■				
John A. Wheeler																			
Herbert F. York								■	■	■	✕	✕							
Gale Young													■						

PROPULSION PANEL

	46	47	48	49	50	51	52	53	54	55	56	57	58	59	60	61	62	63	64
Milton U. Clauser														★					
Allen F. Donovan														✕	✕			✕	✕
Louis G. Dunn						■													
Howard W. Emmons																■	■		
Antonio Ferri																■	■	■	■
Alexander H. Flax																	✕	✕	
William R. Hawthorne						■	■	■											
William M. Holaday						■	■	■											
Abraham Hyatt																■			
Andrew Kalitinsky							■	■											
Mark P. L. Love, Jr.														■	■	■			
Mark M. Mills											✕	✕							
Robert L. Pigford														■	■				
Perry W. Pratt														■	■	■			
William D. Rannie								■	■	■	■								
Addison M. Rothrock																■			
R. E. Schrieber															■				
Ascher Shapiro												■	■						
Abraham Silverstein												■							
C. Branson Smith																			
C. Richard Soderberg					✕	✕	✕	✕	✕										
Homer J. Stewart															■	■			
George P. Sutton																■	■	■	
William J. Sweeney					■														
Edward S. Taylor											■	■	■	■	■				
Phillip B. Taylor							■												
Frank L. Wattendorf						■			■	■									
Gale Young																			

APPENDIX E (Continued)
PSYCH AND SOCIAL SCIENCES PANEL

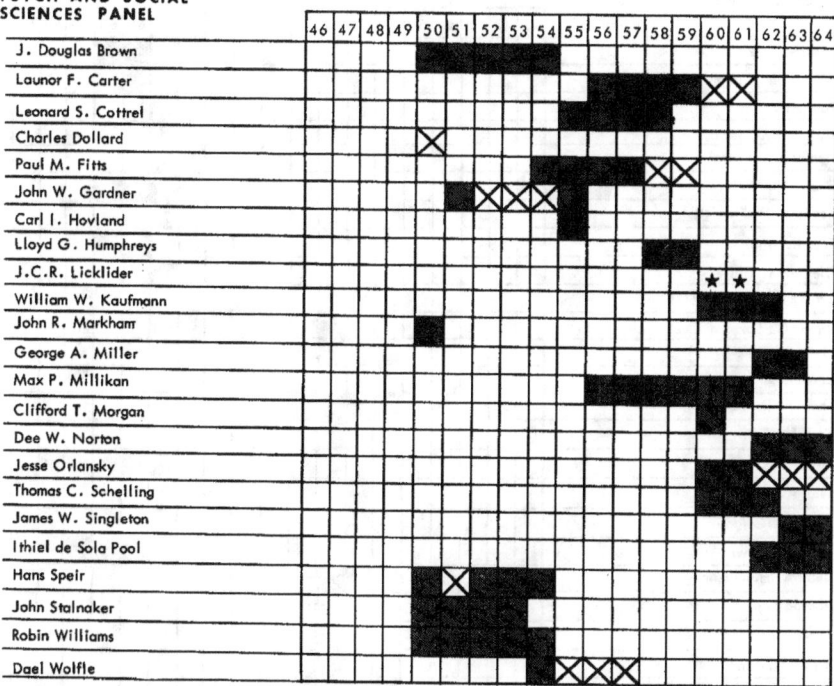

Name	46	47	48	49	50	51	52	53	54	55	56	57	58	59	60	61	62	63	64
J. Douglas Brown					■	■	■	■	■										
Launor F. Carter										■	■	■	■		×	×			
Leonard S. Cottrel																			
Charles Dollard					×														
Paul M. Fitts									■	■	■	■		×	×				
John W. Gardner						■	×	×	×	■									
Carl I. Hovland										■									
Lloyd G. Humphreys													■	■					
J.C.R. Licklider														★	★				
William W. Kaufmann															■	■			
John R. Markham						■													
George A. Miller																	■	■	
Max P. Millikan											■	■	■	■					
Clifford T. Morgan										■	■	■							
Dee W. Norton															■	■			
Jesse Orlansky												■	■	■		×	×	×	
Thomas C. Schelling													■	■					
James W. Singleton														■	■				
Ithiel de Sola Pool													■	■					
Hans Speir					■	×	■	■	■										
John Stalnaker					■	■	■	■		■									
Robin Williams					■	■	■	■											
Dael Wolfle										×	×	×							

RECONNAISSANCE PANEL

Name	46	47	48	49	50	51	52	53	54	55	56	57	58	59	60	61	62	63	64
James G. Baker								■	×	×	■	■							
Allen F. Donovan								★	★	★	★	★							
George B. Kistiakowsky								×											
Edwin H. Land								■	■	■	■	■							
J.C.R. Licklider															★				
Duncan E. Macdonald										■	■	×	×	×	×				
Stewart E. Miller										■									
Brian O'Brien													■	■		×	×		
Carl F. J. Overhage								■	■	×	×	■							
Milton S. Plesset									■	■	■								
William R. Rambo											■	■	■	■					
Richard Raymond											■	■	■						
Herbert S. Scoville, Jr.										■	■	■							
Philip G. Strong										■	■	■							

SPACE TECHNOLOGY PANEL

Name	46	47	48	49	50	51	52	53	54	55	56	57	58	59	60	61	62	63	64
Edward J. Barlow														■	■				
Launor F. Carter														■	■				
David T. Griggs														■	■				
S. W. Herwald														■	■				
Duncan E. Macdonald														■	■				
John P. Marbarger														■	■				
John R. Moore														■	■				
Courtland D. Perkins															×				
Ernst H. Plesset														■	■				

APPENDIX E (Continued)
SPACE TECHNOLOGY PANEL
(Continued)

	46	47	48	49	50	51	52	53	54	55	56	57	58	59	60	61	62	63	64
Perry W. Pratt														X	X				
William H. Radford														X	X				
Paul A. Smith														X					
H. Guyford Stever														X	X				
Homer J. Stewart																			
George E. Valley, Jr.														X	X				
Teddy F. Walkowicz														X	X				
Fred L. Whipple														X	X				
Clayton S. White														X					

MEMBERS-AT-LARGE

	46	47	48	49	50	51	52	53	54	55	56	57	58	59	60	61	62	63	64
James G. Baker							X	X											
Detlev W. Bronk						X	X												
Launor F. Carter																	X		
Hugh L. Dryden						X	X	X											
Paul M. Fitts							X	X											
Alexander H. Flax														X	X				
Ivan A. Getting							X												
David T. Griggs							X												
Richard Horner														X	X				
John A. Hutcheson						X	X												
Joseph Kaplan																			
George B. Kistiakowsky								X											
Edwin H. Land																			
Thomas G. Lanphier, Jr.						X													
W. Randolph Lovelace, II							X												
John R. Markham						X	X												
James McCormack														X	X	X		X	X
Clark B. Millikan							X												
Carl F. J. Overhage								X											
Milton S. Plesset																			
Donald A. Quarles						X													
Louis N. Ridenour, Jr.						X	X												
Leonard S. Sheingold																X			
John von Neumann							X												
Edward Teller						X													
George E. Valley, Jr.								X											
Frank L. Wattendorf						X	X												

ASSOCIATE ADVISORS

	46	47	48	49	50	51	52	53	54	55	56	57	58	59	60	61	62	63	64
James B. Edson																		X	X
Alfred J. Eggers, Jr.																			X
Robert F. Garbarini																		X	X
Abraham Hyatt																		X	
J.C.R. Licklider																		X	X
Donald A. Rice																		X	X
Robert C. Seamans, Jr.																		X	X
Robert J. Shank																		X	X
Alan H. Shapley																		X	X
Frank L. Wattendorf																		X	X
Albert D. Wheelon																		X	X
Fred L. Whipple																		X	X
Robert B. Young																		X	

APPENDIX F
SAB STEERING AND EXECUTIVE COMMITTEES*

	46	47	48	49	50	51	52	53	54	55	56	57	58	59	60	61	62	63	64
STEERING COMMITTEE																			
Robert W. Buchheim																			
Launor F. Carter																			
L/Gen James Ferguson																			
Alexander H. Flax																			
Winston R. Markey																			
Brockway McMillan																			
Courtland D. Perkins																			
Leonard S. Sheingold																			
H. Guyford Stever																			
EXECUTIVE COMMITEE																			
L/Gen Samuel E. Anderson																			
James G. Baker																			
Edward J. Barlow																			
Seymour Bogdonoff																			
Robert W. Buchheim																			
Loren D. Carlson																			
Launor F. Carter																			
Joseph V. Charyk																			
L/Gen Laurence C. Craigie																			
M/Gen Marvin C. Demler																			
Gerald P. Dinneen																			
Charles Dollard																			
James H. Doolittle																			
Charles S. Draper																			
Hugh L. Dryden																			
Lee A. Dubridge																			
Allen F. Donovan																			
L/Gen James Ferguson																			
Enrico Fermi																			
Paul M. Fitts																			
Alexander H. Flax																			
John W. Gardner																			
Ivan A. Getting																			
Ralph E. Gibson																			
David T. Grig																			
Donald W. Hastings																			
M/Gen Victor R. Haugen																			
S. W. Herwald																			
John B. Hickam																			
Henry G. Houghton																			
Joseph Kaplan																			
William W. Kellogg																			
Mervin J. Kelly																			
Robert H. Kent																			
George B. Kistiakowsky																			

* ▬ — Years of membership.

153

APPENDIX F (Continued)

EXECUTIVE COMMITEE

Name	46	47	48	49	50	51	52	53	54	55	56	57	58	59	60	61	62	63	64
Albert L. Latter																			■
W. Randolph Lovelace, II	■	■	■	■	■	■													
Duncan E. Macdonald												■	■	■					
Winston R. Markey																			
John R. Markham					■														
James McCormack																■	■	■	
Gerald M. McDonnel																	■	■	■
William G. McMillan																	■	■	
Clark B. Millikan								■	■	■	■								
Mark M. Mills										■	■	■							
Brian O'Brien														■	■	■			
Jesse Orlansky														■	■	■			
Carl F. J. Overhage										■					■	■			
Courtland D. Perkins													■	■	■	■	■		
Ernst H. Plesset												■	■	■					
Gen Thomas S. Power											■								
Donald L. Putt				■	■	■													
William H. Radford																■	■		
Louis N. Ridenour				■															
L. Eugene Root						■													
Walter A. Rosenblith															■	■			
Thomas C. Schelling																			
Edgar Schmued												■	■	■					
Gen Bernard A. Schriever													■	■					
William R. Sears				■															
Leonard S. Sheingold																			
Abraham Silverstein													■						
Paul A. Smith														■	■	■	■	■	
C. Richard Soderberg	■	■	■	■	■	■													
Hans Speir																			
H. Guyford Stever												■	■	■	■	■	■	■	
Homer J. Stewart						■													
Julius A. Stratton	■																		
M/Gen Leland S. Stranathan										■	■	■							
M/Gen Ralph P. Swofford										■	■								
Edward Teller												■	■	■	■	■	■	■	
George E. Valley, Jr.								■	■	■									
Theodore H. von Karman	■	■	■	■	■	■	■	■	■	■	■	■	■	■					
John von Neumann								■	■										
Shields Warren														■	■	■			
Frank L. Wattendorf								■											
Harry Wexler								■	■										
Clayton S. White												■	■	■					
Fred L. Whipple													■	■					
L/Gen Roscoe C. Wilson														■	■	■			
Dael Wolfle		■						■	■	■									
Vladimir K. Zworykin																			
Total Membership: Executive Committee	6	6	8	8	12	12	13	12	13	13	17	17	21	23	26	25	27	26	25

SAB MEMBERSHIP ON DIVISION ADVISORY GROUPS *

	1962	1963	1964
AERONAUTICAL SYSTEMS DIVISION			
Holt Ashley	■	■	■
Edward J. Barlow	■		
Seymour Bogdonoff		✕	✕
Alexander H. Flax	✕		
J. C. Fletcher			■
Don Flickinger		■	
William W. Kellogg	■		
George P. Sutton			■
Walter W. Wrigley		■	■
BALLISTIC SYSTEMS DIVISION			
Charles S. Draper	■	■	■
Antonio Ferri		■	■
John S. Foster, Jr.	■		
David T. Griggs	■		✕
Albert L. Latter		■	
Conrad Longmire	■	■	
William G. McMillan		■	■
Walter A. Rosenblith	■		
Edward Teller	✕	✕	
ELECTRONIC SYSTEMS DIVISION			
Bruce Billings	■	■	
Gerald P. Dinneen			■
J. C. Fletcher			■
George A. Miller	■		
James W. Meyer	■		
Brian O'Brien	■	■	
William H. Radford	✕	■	
Leonard S. Sheingold		✕	✕
George E. Valley, Jr.	■	■	
Willis H. Ware			■
FOREIGN TECHNOLOGY DIVISION			
Seymour Bogdonoff	■		
Gerald P. Dinneen	■	■	
Allen F. Donovan	■	■	■
S. W. Herwald			■
Richard E. Horner			■
Duncan E. Macdonald	■	■	
Brian O'Brien	✕	✕	✕
Teddy F. Walkowicz	■		

* ■ Member.

✕ Chairman.

APPENDIX G (Continued)

	1962	1963	1964
RANGE TECHNICAL ADVISORY GROUP			
Launor F. Carter	■		
S. W. Herwald	■	✕	✕
Albert L. Latter	■		
William G. McMillan	■		
Brian O'Brien	■	■	■
Leonard S. Sheingold	✕	■	■
Chalmers W. Sherwin	■	■	■
William H. Radford	■	■	■
SPACE SYSTEMS DIVISION			
Nicholas J. Hoff	■		
Robert E. Holzer	■		
Gerald M. McDonnel	■	■	■
Clark B. Millikan	✕	✕	✕
Earle E. Partridge		■	■
Ernst H. Plesset	■	■	■
Homer J. Stewart	■	■	■
Robert N. Thorn	■		

APPENDIX H
GENERAL BOARD MEETINGS
USAF SCIENTIFIC ADVISORY BOARD
1946-1964

Date		Location	Subject
1946	16-20 Jun	Pentagon and Wright Field, Ohio	AAF R&D programs
1947	4-5 Feb	Pentagon	AAF R&D programs
1948	17-18 Mar	Pentagon	USAF R&D programs
	16-18 Nov	Pentagon and Wright Field	USAF R&D programs
1949	5-7 Apr	Pentagon	USAF R&D programs
	2-3 Nov	Pentagon	Reorganization of USAF R&D
1950	12-13 Apr	Pentagon	USAF R&D programs
	11-15 Sep	Pentagon	USAF preparedness in light of problems posed by war in Korea
1951	9-10 May	Pentagon	USAF experience and operational-materiel problems in Korean War
1952	21-23 Jan	AF Missile Test Center, Patrick AFB, Cocoa, Fla.	Long-range USAF R&D plans
	9-11 Sep	AF Cambridge Research Center, Cambridge, Mass.	USAF R&D programs
1953	29-31 Mar	Maxwell AFB, Ala.	USAF R&D programs
	18-22 Oct	Ent AFB, Colorado Springs, Colo.	Air defense
1954	19-21 Mar	Hotel Chamberlin, Old Point Comfort, Va.	Tactical aviation
	27-29 Sep	Offutt AFB, Omaha, Nebr.	Strategic air
1955	24-26 Mar	RAND Corp., Santa Monica Calif.	RAND programs and procedures
	13-15 Jun	Madden Lodge, Brainerd, Minn.	"Think" session
	19-21 Oct	Pentagon	USAF R&D programs

Date	Location	Subject
1956 11-12 Apr	AF Armament Center, Elgin AFB, Fla.	USAF R&D programs
14-16 Nov	San Marcos Hotel, Chandler, Ariz.	"Think" session
1957 20-22 May	Patrick AFB, Cocoa Beach Fla.	USAF R&D programs
4-6 Dec	San Marcos Hotel, Chandler, Ariz.	"Think" session; Limited Warfare
1958 30 Apr-2 May	Wright Patterson AFB, Dayton, Ohio	AMC and Wright Air Development Center programs
22-24 Oct	Ramey AFB, Puerto Rico	"Think" session; Electronics
1959 20-22 Apr	Miramar Hotel, Santa Monica, Calif., with field trip to Vandenberg AFB, Calif.	Guided missiles and space technology
26-28 Oct	Orlando AFB, Fla.	USAF R&D programs
1960 20-22 Apr	AF Academy, Colorado Springs, Colo.	Aerospace defense
24-26 Oct	L. G. Hanscom Field, Bedford, Mass.	USAF command and control requirements
1961 26-28 Apr	San Carlos Hotel, U. S. Navy Postgraduate School, Monterey, Calif.	Arms control and advanced ideas in science
6-8 Nov	San Marcos Hotel, Chandler, Ariz.	National space program
1962 24-26 Apr	Kirtland AFB, Albuquerque, N. Mex.	Limited war; counterinsurgency
25-26 Oct	School of Aerospace Medicine, Brooks AFB, San Antonio, Tex.	Manned space flight requirements
1963 11-12 Apr	Space Systems Division (Arbor Vitae Complex), Inglewood, Calif.	USAF R&D long-range Plans and requirements
24-25 Oct	Offutt AFB, Omaha, Nebr.	Strategic air
1964 27-28 Apr	Eglin AFB, Fla.	Tactical air capabilities
28-29 Oct	Naval Training Center, San Diego, Calif.	Navigation

APPENDIX I

CHANGES IN AIR FORCE REGULATION 20-30*

1948-1964

Subject	14 May 48 (First Issued)	25 Jul 51 (Amended)	2 Jul 56 (Revised)	17 Apr 57 (Amended)	17 Sep 59 (Revised)	23 Nov 62 (Revised)
To Whom Board Reports	CSAF	Same	Same	Contained only one revision: to better inform USAF activities of SAB's services. Informed that board was prepared to consider problems and furnish advice as well as provide consultant services.	Same	Revision: SAF and CSAF.
Composition of Board						Subject to periodic review and approval by SAF and CSAF.
Appointment of Chairman (and Vice Chairman)	Appointed by CSAF for indefinite terms	Same	Addition: To be civilians not in full-time employment of any military department. Appointments to end automatically on departure of CSAF unless new CSAF reappoints them.		Same	Revision: Recommended and appointed by CofS USAF with the approval of SecAF (for indefinite terms).

AFR 20-30 is the official directive which establishes, defines, and explains the mission and organization of the SAB. It was updated three times and amended twice since first published in May 1948.

APPENDIX I

Subject	14 May 48 (First Issued)	25 Jul 51 (Amended)	2 Jul 56 (Revised)	17 Apr 57 (Amended)	17 Sep 59 (Revised)	23 Nov 62 (Revised)
Military Director	To be Dir R&D Hq USAF. To report directly to CSAF on board matters	*Revision:* To be a general officer on Hq USAF staff. Appointed by CSAF after consultation with SAB chairman. To report directly to CSAF on board matters	Same		Same	*Revision:* To be DCS/Research and Development, Hq USAF (and to report directly to CSAF on board matters).
Secretariat	To be appointed by military director with approval of chairman.	*Revision:* A permanent Secretariat will be appointed....	Same		Same	Same
Members	Appointed by CSAF (on recommendation of chairman). Length of appointment at discretion of CSAF and chairman.	Same	*Addition:* Will be granted the same courtesies as lt generals.	Same	Same	*Revision:* Appointed by CSAF (on recommendation of chairman) with approval of SAF.

APPENDIX I

Subject	14 May 48 (First Issued)	25 Jul 51 (Amended)	2 Jul 56 (Revised)	17 Apr 57 (Amended)	17 Sep 59 (Revised)	23 Nov 62 (Revised)
Ex-Officio Members	Military director to be ex-officio member.	*Revision:* Dir R&D, Hq USAF, only ex-officio member.	*Revision:* Membership expanded to include DCS/Development; Chief Scientist; Dir R&D; Dir Dev Planning; Commander ARDC; others as designated by CSAF.	Same	*Revision:* Subject deleted from regulation because it encouraged officers, who failed to understand its intent, to apply for board membership	Same
Guest Members and Consultants			(Practice in vogue since start of board but never before included in regulation) May be invited to serve as appropriate. Accorded full privileges while serving.	Same	Same	Same

APPENDIX I

Subject	14 May 48 (First Issued)	25 Jul 51 (Amended)	2 Jul 56 (Revised)	17 Apr 57 (Amended)	17 Sep 59 (Revised)	23 Nov 62 (Revised)
Executive Committee	Composed of chairman, vice chairman, military director, and panel chairmen	*Additions:* ... and Dir R&D, Hq USAF (as ex-officio SAB member)	*Additions:* ... and Chief Scientist, DCS/Develop, Dir Develop Planning, Commander ARDC.	Same	Same	*Revision:* Composed of chairman, vice chairman, military director, panel chairmen, Asst SAF (R&D), senior statesmen, Chief Scientist, Commander AFSC; (others as appropriate).
Steering Committee						Composed of chairman, vice chairman, military director, Asst SAF (R&D), Chief Scientist.
General Board Meetings	To be held at times agreed on by CSAF, chairman, and military director.	Same	*Revision:* To be held usually twice a year on call of chairman.	Same	Same	Same

APPENDIX I

Subject	14 May 48 (First Issued)	25 Jul 51 (Amended)	2 Jul 56 (Revised)	17 Apr 57 (Amended)	17 Sep 59 (Revised)	23 Nov 62 (Revised)
Implementation of Board Recommendations (Reports, etc)		Unless CSAF orders otherwise, DCS/Dev will monitor board recommendations. These will be implemented or actions taken on reasons for not doing so will be sent to CSAF. Chairman of board will be notified in writing of all all recommendations.	*Revision:* The requirements to inform CSAF when board recommendations were not implemented and to inform the chairman in writing of actions taken on board recommendations were re-inserted from regulation into Hq USAF HOI.	Same	Same	*Revision:* DCS/Research and Development to monitor implementation of board recommendations and (in his capacity as military director) to determine the distribution of these reports.

APPENDIX J

STUDIES AND AUTHORS*
TOWARD NEW HORIZONS
Dec 15, 1945

STUDY	*AUTHOR*

SCIENCE, THE KEY TO AIR SUPREMACY ... Dr. Theodore von Karman

WHERE WE STAND ....Dr. von Karman

AERODYNAMICS AND AIRCRAFT DESIGN

Part I: High Speed Aerodynamics ....Dr. Hsue-shen Tsien

Part II: The Airplane-Prospects and
          Problems ....Dr. William R. Sears
                                     Mr. Irving L. Ashkenas
                                     Capt. Chester N. Hasert

Part III: Aircraft Materials and Struc-
          tures ....Dr. Nathan M. Newmark

FUTURE AIRBORNE ARMIES.............Maj. Teddy F. Walkowicz

AIRCRAFT POWER PLANTS

Part I: Gas Turbine Propulsion ......Dr. Frank L. Wattendorf

Part II: Experimental and Theoretical
          Performance of Aeropulse
          Engines ....Dr. Tsien

Part III: Performance of Ramjets and
          their Design Problems ......Dr. Tsien

Part IV: Future Trends in the Design
          and Development of Solid and
          Liquid Fuel Rockets ....Dr. Tsien

Part V: High Temperature Materials ....Dr. Pol Duwez

AIRCRAFT FUELS AND PROPELLANTS

Part I: Research on Hydrocarbon Fuels
          for Aircraft Propulsion .....Dr. William J. Sweeney

Part II: Petroleum: Its Use for Motive
          Power ....Dr. Sweeney

Part III: Solid Propellants for Rockets and
          Other Jet Propelled Devices..Dr. Louis P. Hammett

---

*Other civilian contributors to the work of the Scientific Advisory Group whose names did not appear on the studies were: Mr. Morton Alperin (Technical-Ass't) ; Mr. Stanley N. Brower; Dr. W. S. Hunter; Prof. John R. Markham; Mr. Henry Nagamatsu; Dr. Ernest E. Sechler; Mr. George S. Schairer; Dr. Theodore H. Troller; Dr. Fritz Zwicky; and Dr. Vladimir K. Zworykin. Military officers who served with the group in administrative capacities were: Capt Thomas E. Daley; Lt Col Adolph P. Gagge; Col Frederic E. Glantzberg (Military Deputy); Maj Oliver W. Hammonds; Capt Charles H. Jackson, Jr.; Lt Col Godfrey T. McHugh; Capt McGowan Miller; Lt Col Frank W. Williams; and Col Donald N. Yates.

APPENDIX K

REGISTER OF SAB REPORTS*

AD HOC AND SPECIAL COMMITTEE REPORTS

Date	Subject/Members
21 Sep 49	Research and development in the USAF (*Ridenour*, G. P. Baker, Doolittle, J. B. Fisk, Overhage, R. A. Sawyer, Wattendorf, J. M. Wild, R. J. Woodrow).
Nov 49	Site for Arnold Engineering Development Center (*Markham*, Hoff, Lovelace, Stever, Wattendorf).
21 Feb 50	Plans, program, and progress of Project RAND (*DuBridge*, Duwez, Kaplan, Ridenour, Robertson, Sears).
3 Apr 50	Annual report of SAB Project RAND Committee (*Gibson*, Duwez, Kaplan, Ridenour, Robertson, Sears)
26 Apr 50	Recommendations for operation of Arnold Engineering Development Center (*Markham*, Berlin, Bollay, Wattendorf).
1 May 50	Progress Report: Air Defense Systems Engineering Committee (*Valley*, Draper, Donovan, Houghton, Stever).
24 May 50	Electronic and control system proposals for the 1954 interceptor (Astin, Donovan, Stever, Valley).
Jul 50	Recommendations on USAF guided missiles program (*Ridenour*, Astin, F. H. Clauser, Donovan, Stever)
Dec 50	Medical research and development in the USAF (*Lovelace*, Baldes, Carlson, R. L. Clark, Fitts, M. I. Gregerson, Hastings, Hickam, Kaplan, J. H. Laurence, Warren).
7 Aug 51	Min of Mtg of Special Working Gp of SAB on mobilization of scientific personnel in support of USAF R&D (*Ridenour*, H. F. Bohnenblust, F. H. Clauser, Fisk, Getting Overhage, R. J. Woodrow).
4 Oct 51	USAF armament activities and requirements (Hutcheson, Draper, H. T. Hokanson, Kent, Lauritsen, Schmued, L. T. E. Thompson).
26 Aug 52	USAF electronic countermeasures (ECM) Quick Reaction facilities and procedures (*Pollard*, et al).
4 Dec 52	Survey of new developments and applications in materials in European research laboratories (Duwez)
13 Dec 52	Recommendations on future development of Project ATLAS program (*C. B. Millikan*, Bode, M. U. Clauser, Draper, Kistiakowsky, G. F. Metcalf, Stewart, M. J. Zucrow).
1 Sep 53	Aircraft nuclear propulsion program (*Soderberg*, Donovan, Griggs, Hastings, Holaday, Markham, C. B. Millikan, Mills, Rannie, Rothrock, Stewart, Teller, Warren, G. Young).

---

*Names of committee chairmen are in italics. SAB members are noted by their surnames only except for those with like names; otherwise, the person noted served in a consultant capacity.

Ad Hoc and Special Committee Reports—continued

Date   Subject/Members

Aug 53   Development and procurement of USAF training equipment (Wolfle and R. L. Garman).

19 Oct 54   Nuclear missile propulsion (*Mills*, Bradbury, Donovan, Dunn, Duwez, D. Froman, C. B. Millikan, Rannie, Rothrock, Silverstein, Teller, von Neumann, York, G. Young).

8 Feb 55   Nuclear missile propulsion (see membership above, 19 Oct 54).

14 May 55   Aircraft nuclear propulsion (*C. B. Millikan*, Donovan, Mills, Rannie, Root, Rothrock, Schairer, Stever, E. Taylor, Teller, Warren, R. Widmer, G. Young).

31 May 55   Ballistic missiles defense (*Stever*, Barlow, Bode, D. D. Dustin, Getting, Gilruth, W. Graham, A. G. Hill, C. A. Lindbergh, A. Longacre, E. A. Martinelli, R. F. Mettler, S. E. Miller, W. K. H. Panofsky, Sherwin, Stewart, York).

15 Jun 55   Aircraft nuclear propulsion (see membership above, 14 May 55).

28 Jun 55   Coherent focused side-looking doppler radar (Sherwin).

Feb 56   Nuclear Propulsion of missiles (see membership above, 19 Oct. 54 and 8 Feb 55).

10 Apr 56   Review and concurrence in Air Technical Intelligence Center proposal for closer liaison with SAB (SAB Executive Committee).

Apr 56   FPS-17 program (*Purcell*, R. P. Johnson, S. E. Miller, Sherwin).

25 Jul 56   MB-1 rocket follow-on program (*Root*, Draper, Foster, G. A. Fowler, Lauritsen, A. E. Puckett, Ramo, Stever).

17 Aug 56   Ballistic missiles defense (*Stever*, Barlow, P. Bechtel, Gilruth, W. B. Graham, Holloway, C. A. Lindbergh, Longacre, E. A. Martinelli, R. F. Mettler, W. K. H. Panofsky, Stewart, A. G. Wimer.

23 Aug 56   Joint development of ballistics range (*Sherwin*, Gilruth, E. H. Plesset, Schmued).

14 Mar 57   Solar energy (*Duwez*, F. Daniels, Haurwitz, H. C. Hattel, L. J. Heidt, M. Kastens, G. L. Pearson, Stever, von Karman, A. Zarem).

9 Oct 57   Advanced weapons technology and environment (*Stever*, Kaplan, C. B. Millikan, Mills, Radford, Ramo, White).

15 Oct 57   Base hardening (*E. H. Plesset*, H. N. Bleich, E. Doll, R. J. Hansen, H. Kahn, M. M. May, Newmark, R. B. Peck, Peter, P. O. Weidlinger, White).

6 Dec 57   Space Technology (Griggs, C. B. Millikan, Mills, Radford, Stever, Teller, White).

10 Mar 58   Rotating engine development (*Mills*, Charyk, Love, Perkins, Silverstein).

Ad Hoc and Special Committee Reports—continued

Date	*Subject/Members*
Jun 58	Research and development in USAF (*Stever*, B. Archambault, Lovelace, Morgan, Perkins, Pratt, Sawyer, Walkowicz, R. J. Woodrow).
25 Oct 58	High intensity electric arc-jets (*Duwez*, A. Buseman, M. U. Clauser).
15 Jan 59	R&D requirements for air defense systems (*Sherwin*, Carter, Love, F. A. Payne, Radford, R. Raymond, Ridenour, Schmued, M. Stern).
Mar 59	Limited war (*Ridenour*, A. Androskey, S. T. Cohen, D. Kybal, M. P. Maier, McCormack, M. F. Millikan, E. H. Plesset, R. C. Raymond, P. J. Schenk, Sherwin, Walkowicz, Weiss, D. D. Whitcraft.
17 Jul 59	Aircraft nuclear propulsion program (*Perkins*, Duwez, Harris, D. Novik, E. H. Plesset, Starr).
17 Sep 59	Infrared countermeasures for the B-70 (*Valley*, C. W. Miller, S. Passman, R. H. Rose, E. S. Rubin, Sheingold, Siegel, Townes).
12 Oct 59	High power microwave facility (*L. R. Hafstad*, S. C. Brown, Ferri, N. Marcuvitz, O'Brien, A. Peterson, Radford, J. R. Ragazzini, Sherwin, C. Van Atta).
4 Nov 59	USAF radiobiology program (*Tullis*, H. L. Andrews, V. P. Bond, H. D. Bruner, C. L. Dunham, L. H. Hempelmann, J. L. Liverman).
31 Dec 59	Nuclear testing (*Doolittle*, Agnew, Carter, S. T. Cohen, Flax, Foster, Griggs, Lauritsen, McCormack, E. H. Plesset, P. A. Smith, Stever, Teller, J. Wheeler, White, M. Zelle).
Mar 60	Supplementary findings to SAB Ad Hoc Committee report on radiation facility (see membership above, 12 Oct 59).
20 Jun 60	Very low frequency radar (*Radford*, J. H. Chisholm, G. C. Comstock, C. M. Crain, Siegel, Weiss, Villard).
Jul 60	Aircraft nuclear propulsion program (*E. H. Plesset*, A. T. Biehl, D. D. Corson, Duwez, Harris, J. H. Jackson, D. Novik, F. A. Payne, Starr, R. H. Widmer, C. Wood).
17 Aug 60	Sensors for space surveillance (Herwald, Fubini, W. Graham, W. Huggins, R. Muchmore, J. W. Schaefer, W. M. Siebert, P. A. Smith, Whipple, Wrigley).
Sep 60	USAF occupational health (toxicological) programs (*Princi*, Buchan, K. P. Dubois, Tullis).
Nov 60	Space counter weapon systems (*Stever*, Ashley, L. Goldberg, W. Graham, Herwald, R. H. Kingston, A. Latter, Marbarger, R. Miller, E. H. Plesset, Pratt, Radford, E. Rubin, Schilling, Sheingold, Siegel, Valley, Whipple, Wrigley).

Ad Hoc and Special Committee Reports—continued

Date	*Subject/Members*
Dec 60	Energy conversion for space power applications-exclusive of propulsion (*E. H. Plesset, A. T. Biehl* (co-chairmen), C. M. Kelly, L. J. Koch, J. A. Krumhansl, Merkle, W. B. Nottingham, W. C. Scott, U. B. Thomas, C. Tobias, E. B. Yeager).
Dec 60	Aerospace plane (Vehicles, Propulsion, and Space Technology panel members)
Jan 61	ARDC wind tunnel facility (*Bogdonoff*, Ferri, Flax, C. B. Millikan).
May 61	Utilization of scientific resources (*Stever*, Horner, McCormack, C. B. Millikan, Perkins, Putt, Valley).
6 Jun 61	Space surveillance: review of August 1960 SAB report on subject (*Flax*, Bogdonoff, Donovan, Eggers, Emmons, Hoff, Perkins, Pigford, Schreiber, Stewart).
16 Jun 61	Neutralization of Soviet hardened sites (*Peter*, H. L. Brode, A. C. Haussmann, Healey, A. H. Katz, Kellogg, Newmark, R. H. Shatz, R. M. Smith).
14 Jul 61	Manned military space program (*Perkins*, Bogdonoff, J. Bollerud, Flax, Hoff, J. Irving, Sheingold).
31 Jul 61	Assessment of possible military applications of ionospheric modifications (*Holzer*, A. J. Dressler, L. Goldberg, Kellogg, Shapley, Siegel, Villard, Wheelon, A. J. Zmuda).
Aug 61	Protective structures applied research program (*McMillan*, M. S. Agababian, H. L. Brode, Cook, R. E. Fadum, E. A. Martinelli, C. Violet, A. E. Ward, W. M. Wells).
Aug 61	Aerospace plane (*Flax*, Bogdonoff, Donovan, Eggers, Emmons, Perkins, Pigford).
Oct 61	Life sciences-human factors facilities relative to human centrifuges, national resources and needs (Interim Report) (Buchan, Carlson, Marbarger, Orlansky, Singleton, Wagner).
Oct 61	Passive satellite communications (*B. F. Miller*, P. Pontecorvo, G. Raisbeck, S. Reiger, Tinus, Wheelon).
Oct 61	Manned military space program (*Perkins*, Bogdonoff, Flax, Hoff, McDonnel, Norton, Sheingold).
Oct 61	Project Sky Bolt (*Perkins*, Flax, Healey, H. Lawrence, Radford, Shank).
Dec 61	AFSC technical facilities acquisition procedures and authority (*Horner*, Ashley, Buchan, Ferri, Flax, R. F. Garbarini, Macdonald, Marbarger, Meyer, Pratt, Reed, B. F. Ruffner, P. A. Smith).
Jan 62	Mathematical Biological models for manned space operations (*Adey*, R. Bellman, J. Bigelow, V. Bond, Carlson, Lion, G. Miller, Ware).
Feb 62	AFSC technical facilities (follow-up on December 1961 report).

Ad Hoc and Special Committee Reports—continued

Date	Subject/Members

Apr 62   Effectiveness of USAF in-house laboratories organization and manning (*Sheingold,* Carter, P. V. Cusick, Flax, Healey, Herwald, Horner, Licklider, Orlansky, Pratt, Radford, Schmued, P. A. Smith, Valley, R. C. Wilson).

Jul 62   Aerospaceplane program (*Perkins,* Barlow, R. Bussard, Eggers, Hoff, R. H. Miller, Pigford, Stewart).

Sep 62   XB-70 lift-to-drag ratio (Flax, R. H. Widmer, J. K. Wimpress).

Sep 62   USAF space program (*Stever,* R. W. Buchheim, Carter, Flax, Getting, Griggs, Herwald, McDonnel, C. B. Millikan, Perkins, T. B. Taylor).

Nov 62   Nuclear propulsion technology (*E. H. Plesset,* Duwez, C. E. Clifford, Flax, C. T. Leondes, McDonnel, Perkins, Porter, Pratt, Starr, Thorn).

Dec 62   Machine language translation (*Dinneen,* F. W. Householder, Madden, G. A. Miller, Tonge, W. S-Y Wang, Ware).

Dec 62   Strategic concepts (*Doolittle,* M/Gen R. A. Breitweiser, Carter, M/Gen S. J. Donovan, Foster, Flax, Griggs, B/Gen H. A. Hanes, M/Gen W. B. Keese, McDonnel, M/Gen J. D. Page, Partridge, Perkins, L. B. Rumph, G. P. Saville, Valley).

Jun 63   Space radiation effects (*Radford,* J. B. Edson, Kaplan, King, Latter, Meyer, B. F. Miller, O'Brien, Porter, Sheingold, Sherwin, R. A. Smith, Tinus, A. D. Wheelon, C. D. Zerby).

Nov 63   Tactical penetration capabilities (*Fletcher,* R. M. Ashby, R. Kahal, J. P. Haverty, Rambo, Sheingold, Col H. E. Walmer, L/Col W. B. Williamson).

Dec 63   ESD computer technology development plan (*Dinneen,* Orlansky, Tonge, Ware).

Jan 64   RTD laboratories (Latter, Perkins, Radford, Sheingold, P. A. Smith, Stever, Valley).

Apr 64   Reconnaissance and intelligence in counterinsurgency (*Billings,* A. Katz, Macdonald, Sheingold).

28 May 64   Military political science (*Orlansky,* R. W. Buchheim, Carter, J. M. Goldsen).

Jul 64   Boron research (*R. H. Miller,* L. Brewer, R. J. Diefendorf, Duwez, D. L. Grimes, Hoff, A. J. Kullas, W. R. Micks, E. R. Parker, H. Ponsford, P. E. Sandorff).

Aug 64   Hypersonic wind tunnel (TRIPLETEE) facility (*Bogdnoff,* Ferri, C. B. Smith).

Oct 64   Military implications of engineering geology concerning USAF civil engineering research (*Reed,* W. Bailey, D. Carder, J. B. Edson, G. O. Gates, M. Holter, W. Judd, D. Norton, Porter, P. A. Smith).

PANEL REPORTS*

AEROMEDICAL/BIOSCIENCES

Date	*Subject*
19 May 52	Review of aeromedical research within Wright Air Development Center.
29 Dec 52	Evaluation of human factor requirements for new aircraft.
9 Feb 55	Procedures for ensuring adequate supply of scientists for the conduct of aeromedical research within USAF.
28 Mar 55	Review of the radiobiological research program at the USAF School of Aviation Medicine.
Apr 58	Summary of activities: January 1957-April 1958.
28 Apr 58	Review of USAF aeromedical research projects.
Oct 58	Review of USAF School of Aviation Medicine activities.
12 Jan 60	Bioinstrumentation
29 Mar 60	Results of joint meeting with USAF Medical Research Council.
Jan 61	USAF life sciences program.
Jun 61	USSR/NASA/USAF bioastronautics programs.
Nov 62	USAF bioastronautics facilities, bioinstrumentation, and radiobiology.
May 63	Advanced Bioastronautics considerations.
Dec 63	Report on 1963 activities.
Oct 64	Report on 1964 activities.

AEROSPACE VEHICLES

11 Sep 51	Long range plans for strategic bombing systems.
28 Sep 51	Turbo-prop power plants (jointly with Propulsion Panel).
Nov 54	AVRO project Y2 aircraft (jointly with Propulsion Panel).
18 Jan 55	Successor to B-52 weapon system.
8 Feb 55	Boundary layer control, extreme altitude reconnaissance aircraft, and cooperative research airplane program.
12 Apr 56	Research aircraft and airframes (jointly with Propulsion Panel).
20 Aug 56	AVRO project.
May 58	Technical status of the B-70 weapon system; recommendations on Dyna Soar and other aerospace vehicles development projects.
18 Dec 58	B-70, F-108, and Dyna Soar projects.
26 Jan 59	Deficiencies in BOMARC test program.
25 Jun 59	Boundary layer control aircraft.
Oct 59	Vehicle fatigue problems.
Jan 60	Dyna Soar project.
15 Apr 60	Review of Dyna Soar project.

*See Appendix D for panel names at the time reports were issued. Appendix E gives panel membership at these times.

PANEL REPORTS: AEROSPACE VEHICLES—Continued

Date	*Subject*
Dec 60	Comments on the Dyna Soar program.
Mar 61	Review of current aeronautical development programs.
Nov 61	Laminar flow control.
Apr 62	Dyna Soar panel flutter.
Nov 62	Technical progress and technical design problems of current aeronautical development programs.
Oct 63	Aerospaceplane, VTOL, and strategic manned aircraft programs.

BASIC RESEARCH

May 60	Crystal growing facilities.
Oct 60	Meeting at Aeronautical Research Laboratories.
Feb 61	AFCRL briefing and recommendations.
Jun 61	Facilities, procedures, and support of OSR.

ELECTRONICS

19 Oct 49	IFF development within USAF.
25 Mar 54	Establishment of European research and development laboratory for assisting in creation of European aircraft control and warning system.
14 Mar 55	Infrared components and their application to weapon systems.
28 Apr 55	USAF electronic warfare management programs.
28 Jun 55	Review of Sherwin study of coherent focused sidelooking doppler radar.
3 Nov 55	High-resolution forward-looking airborne radar program.
3 Nov 55	Thermal terrain sensing systems.
Apr 56	Soviet ground to air missile system.
May 58	Basic research programs in support of USAF electronic requirements.
25 Mar 59	Strengthening of USAF electronic support systems.
26 Apr 60	USAF communications problems and requirements.
May 61	USAF radiation weapons program.
Apr 62	Strengthening of USAF electronic warfare program.
Sep 63	Overland airborne radar program.

GEOPHYSICS

19 Jun 51	Geophysics Research Directorate program, AFCRL.
Dec 52	Recommendations stemming from 1952 panel meetings.
12 May 53	Facilities and procedures of Geophysics Research Directorate.
24 Mar 54	Report of meeting at Air Weather Headquarters.
27 Sep 54	Special report, Dr. Whipple, on first steps in using artificial satellites.
12 Apr 56	Applications of geophysical research to weapon systems.
Jun 56	Facilities and support of Geophysics Research Directorate.
28 Feb 57	Geophysical aspects of tropospheric and ionospheric propagation of RF energy.

PANEL REPORTS: GEOPHYSICS—Continued

Date	Subject

6 Aug 57 Significant technical projects and areas in geophysical research.

Feb 59 Management of Cambridge Research Center.

23 Jun 59 Training of personnel in geophysics and space physics.

Dec 59 Dependence of military operations on geophysics research.

Jun 60 Arctic research.

Aug 61 Sacramento Peak solar observatory.

Nov 61 Arctic research.

Feb 62 Geophysical warfare.

Apr 62 Geodetic support for ICBM operations.

Apr 62 BMEWS at Clear, Alaska.

May 62 Emergency communications (through earth).

Jun 62 Meteorology and aerospace environmental services.

Jun 62 Upper atmosphere and space environment.

Jul 62 Report of activities: July 1961-July 1962.

Aug 63 Arctic research (updating 1960-1961 reviews).

Dec 63 Use of BMEWS radar for magnetospheric sounding and for communications experiments based on free electron scatter.

Mar 64 USAF solar system observatory.

Dec 64 Report on 1964 activities.

GUIDANCE AND CONTROL

27 Jan 50 Development of new explosives (and general review of overall USAF armament development programs).

13 Mar 51 Recommendations stemming from meetings on general armament problems and bomber airplane defense.

14 Mar 51 Nuclear weapons for tactical application.

18 Nov 52 Strengthening USAF armament research and development programs.

27 Jan 53 USAF biological and chemical warfare programs.

1 May 53 Strike photography in armament development and application.

3 Jun 53 Falcon missile development.

7 Jun 54 Relationship between fire control systems and ordnance requirements; recommendations on bomber defense.

6 Nov 54 General armament research and development.

16 Nov 56 Lead collision fire control system.

31 Jul 57 Nuclear armament considerations (jointly with Nuclear Panel).

Aug 57 Feasibility of remote-air-battle defense tactics for the Air Defense Command.

Oct 58 Examination of the current and foreseeable state-of-the-art in ballistic missile guidance.

26 Mar 59 Inertial guidance R&D test facility.

Apr 59 Importance of adequate lead time planning for testing of weapon systems.

PANEL REPORTS: GUIDANCE AND CONTROL—*Continued*

Date	Subject
11 May 59	MINUTEMAN guidance system.
23 Jul 59	Testing of advanced and future systems.
19 Aug 59	Adequacy of USAF missile range accuracy techniques.
26 Aug 59	Fingerprint guidance system.
Jul 60	Terminal guidance.
Jul 60	USAF satellite interception requirements.
Oct 60	MINUTEMAN guidance system.
Oct 60	AFMTC central inertial guidance test facility.
Jun 61	Advanced ballistic missile guidance concepts.
Nov 61	Recoverable ballistic test bed for guidance and control subsystems.
Jan 62	Orienting advanced ballistic missile guidance techniques to medium range ballistic missile.
May 62	Space related applied research.
Sep 62	Applied research and advanced technology for space related projects.
Mar 63	ADO-51 all-weather tactical air to surface missile.
Mar 63	Terminal guidance.
Jul 63	Titan III guidance program.
Jul 63	USAF terminal guidance requirements and programs.
Nov 63	Relationship of weapon guidance to target signatures.
May 64	Inertial guidance development and testing.
Jul 64	Guidance error analysis vehicles concept.
Sep 64	Feasibility of multiple independent re-entry vehicles concept.
Dec 64	Multiple independent re-entry vehicles.

INFORMATION PROCESSING

Date	Subject
Mar 63	Standardization problems relating to computer programming of command and control requirements.
Aug 63	Digital computer requirements at Air Force Academy.
Apr 64	Digital computer education.

NUCLEAR

Date	Subject
23 Jun 53	Nuclear carrying capacity of future aircraft.
21 Oct 53	Long-term predictions on nuclear weapons development possibilities.
31 May 54	Panel considerations of large and small nuclear weapons development possibilities.
20 Jul 55	Tactical employment of nuclear weapons.
2 Oct 56	Radiation hazards from nuclear-powered aircraft.
6 Oct 56	USAF aircraft nuclear propulsion program.
19 Nov 56	High-altitude weapons effects.
19 Nov 56	Plutonium contamination hazards.

PANEL REPORTS: NUCLEAR—Continued

Date	Subject
7 Jan 57	USAF nuclear propulsion program.
24 May 57	Weapon kill in air defense.
27 May 57	Appraisal of future nuclear weapon research and development possibilities.
31 Jul 57	Nuclear armament considerations (jointly with Guidance and Control Panel).
12 Sep 57	Estimates on future production of fissile material.
10 Dec 57	Substitute warhead in MB-1.
19 Dec 57	Future production of fissile material.
19 Dec 57	Extent of hazard from accident with nuclear weapon.
19 Jun 58	High-altitude nuclear weapon effects and application.
8 Aug 58	Improving Air Force nuclear weapons capabilities under a test moratorium.
17 Nov 58	Nuclear weapon safety requirements.
22 Nov 58	Missile warheads.
22 Jun 59	Outer space testing of nuclear weapons.
21 Jul 59	Funding of the USAF nuclear environmental test facility.
30 Oct 59	Lethal radius of nuclear weapons against ballistic missiles.
Dec 59	Estimated performance of future nuclear weapons.
7 Apr 60	Nuclear pulsed propulsion.
Nov 60	Nuclear howitzer.
Dec 60	Technical problems associated with underground and space testing of nuclear weapons.
Mar 61	Progress of nuclear pulsed propulsion program in past year.
Apr 61	Pure fusion weapons.
Jun 61	New small nuclear weapons development.
Jul 61	Penetration aids for ballistic missiles.
Aug 61	Weapons fabrication.
Oct 61	Importance of atmospheric testing of weapons.
Oct 61	Proposed atmospheric atomic test plans.
Oct 61	Priority of nuclear weapons tests of primary interest to the USAF.
Nov 61	National Security Council nuclear tests policy.
Jan 62	Russian atomic weapons test.
Jan 62	U. S. weapons testing program.
17 Mar 62	Open Ear report #1.
Oct 62	Updating of review of progress of nuclear pulsed propulsion program.
28 Nov 62	Open Ear report #2.
Jun 63	Consideration of USAF weapons effects program.
28 Jul 63	Open Ear report #3.
Dec 63	Updating of review of USAF weapons effects program.
9 Mar 64	Open Ear report #4.

PANEL REPORTS: NUCLEAR—Continued

Date	*Subject*
18 Mar 64	Open Ear report #5.
Apr 64	Appraisal of new developments in nuclear pulsed propulsion programs.
5 Jun 64	Open Ear report #6.
Jun 64	Review of advances in design of multiple warhead possibilities.
8 Oct 64	Open Ear report #7.
4 Dec 64	Open Ear report #8.
31 Dec 64	Open Ear report #9.

PROPULSION

15 Feb 50	Aircraft engine and liquid rocket power plant R&D.
28 Sep 51	Turbo-prop power plants (jointly with Aerospace Vehicles Panel).
Apr 52	On propulsion briefing by ARDC.
Sep 53	Aircraft nuclear propulsion development program.
Nov 54	AVRO Project Y2 aircraft (jointly with Aerospace Vehicles Panel).
12 Apr 56	Research aircraft and airframes (jointly with Aerospace Vehicles Panel).
14 Nov 56	Conventional and novel engines: propulsion for space operations.
10 Mar 58	Storable liquid and solid rocket propulsion systems.
31 Dec 59	Degree of USAF control required for adequate development of propulsion for military purposes.
Dec 61	Review of USAF propulsion program.
Dec 61	Applied research program in propulsion.

PSYCHOLOGY AND SOCIAL SCIENCES

6 Jun 51	ARDC responsibility for human resources research activities.
14 Dec 51	Survey of USAF program for human resources research.
Dec 52	Observations stemming from meetings with research Advisory Council of the Human Resources Research Institute and the Social Sciences Division and Economics Division of RAND.
6 Jan 53	Basic research to be undertaken by the National Science Foundation.
10 Apr 57	Human factor engineering mission as related to the qualitative superiority of future man-machine weapon systems.
Aug 58	Human factors and ballistic missiles.
Aug 59	Scientific and engineering officers.
Dec 60	Human factors in computerized systems.
Oct 61	Unconventional, limited, and guerrilla warfare.

DISCONTINUED PANELS

Date	Subject
15 Jan 52	*Guided Missiles:* Infrared research and development program.
7 Apr 53	*Physical Sciences:* Functions, procedures, and support of OSR.
3 Aug 53	*Intelligence Systems:* Observations on USAF intelligence requirements and programs.
29 Jan 54	*Intelligence Systems:* Aerial photography by moonlight, balloon reconnaissance by moonlight, covert photography.
28 May 56	*Reconnaissance:* Reconnaissance from satellite vehicles.
28 May 56	*Reconnaissance:* Reconnaissance from high altitude balloons.
Apr 58	*Reconnaissance:* WS-117L program.
8 Apr 59	*Space Technology:* Review of elements of ballistic missile program.
9 Apr 59	*Space Technology:* Space technology problem areas.
25 Jun 59	*Space Technology:* Estimate of new ballistic missile capability.
Dec 59	*Reconnaissance:* Overall USAF reconnaissance requirements and current state of technical development.
29 Apr 60	*Space Technology:* USAF studies on requirements for a lunar base.
30 Dec 60	*Space Technology:* Summary report on space technology.
Jun 61	*Arms Control:* Arms control developments.
Apr 62	*Arms Control:* Implications of observation satellites.

NOTES

CHAPTER ONE

1. Gen H. H. Arnold, *Global Mission*, (Harper & Bros, 1949), p 532.

2. *Ibid.*

3. Memo, Dr. F. L. Wattendorf to author, 7 Nov 65, subj: Comments on the USAF SAB History; Ltr, War Department Air Corps Materiel Div (Wright Field) to Chief of the Air Corps, 6 May 40, subj: Employment —Dr. Theodore von Karman.

4. Ltr, Hq AAF to Office Secy of War, 23 Oct 44, subj: Appointment of Expert Consultant—Theodore von Karman; Memo, Dr. Wattendorf, 7 Nov 65.

5. Memo, Gen Arnold to Dr. von Karman, 7 Nov 44, subj: AAF Long Range Development Program.

6. Memo, L/Gen B. M. Giles, Dep Comdr AAF, to Air Stf, 10 Nov 44, subj: AAF Long Range Development Program—Dr. von Karman; Memo, AC/AS Ops, Commitments, & Rqmts to AC/AS Mgt, 22 Nov 44, subj: AAF Long Range Development Program.

7. Hq AAF HOI 20-76, 1 Dec 44, subj: Organization—The AAF Scientific Advisory Group.

8. Arnold, *Global Mission*, p 533.

9. Min of SAG Mtg, 3 Apr 45.

10. Memo, Col F. E. Glantzberg, SAG Mil Dep to AAF Advsy Coun, 3 Aug 45, subj: Organization and Functions of the Scientific Advisory Group.

11. Memo, B/Gen J. F. Phillips, Hq AAF, to Dir AAF Air Tech Svc Comd, 23 Jun 45, subj: Functions Prescribed for Dr. Theodore H. von Karman's Scientific Advisory Group by the CG, AAF.

12. Min of SAG Mtg, 29 Aug 45.

13. Memo, Dr. Wattendorf to author, 7 Nov 65.

14. Dr. von Karman, *Where We Stand*, p iv.

15. *Ibid.*, p 1.

16. Min of SAG Mtg, 29 Aug 45.

17. *Ibid.*

18. Ltr, Dr. von Karman to Gen Arnold, 15 Dec 45.

19. Memo, Dr. von Karman to Gen Arnold, 20 Dec 45, subj: Establishment of a Permanent Scientific Advisory Group.

20. Dr. von Karman, Science, *Key to Air Supremacy*, p 1.

21. *Ibid.*, pp 2-6.

22. *Ibid.*, p 82.

23. Dr. von Karman to Gen Arnold, 15 Dec 45.

24. Ltr, Gen Arnold to M/Gen E. M. Powers, AC/AS-4, 3 Jan 46.

25. Ltr, L/Gen N. F. Twining, CG AMC, to CG AAF, 8 May 46, subj: Recommendations of Vol I of the Report of the AAF Scientific Advisory Group; Memo, M/Gen Powers to Asst DC/AS R&D, 17 May 46, subj: Recommendations of Vol I of the Report of the AAF Scientific Advisory Group.

26. Memo, M/Gen C. E. LeMay, DC/AS R&D, to AC/AS-4, 14 Jun 46, subj: Recommendations of Vol I of the Report of the AAF Scientific Advisory Group.

27. M/Gen D. L. Putt, SAB Mil Dir., Min of Remarks to Stf Mtg of Engineering Div, Hq AMC, 4 Apr 49.

28. Speech, Mr. J. A. Lang, Jr., Admin Asst to SAF, 10 Apr 64, to 16th Annual Conclave, Arnold Air Society, Denver, Colo., as reprinted in SAFOI Plcy Ltr, Sup 131, May 64.

CHAPTER TWO

1. Memo of Final Mtg of AAF Scientific Group, 6 Feb 46.

2. Ltr, Dr. von Karman to Gen C. A. Spaatz, Actg CG AAF, 27 Feb 46.

3. Dr. von Karman, *Science, Key to Air Supremacy*, p 101.

4. Memo, Dr. von Karman to Gen Arnold, 20 Dec 45, subj: Establishment of a Permanent Scientific Advisory Group.

5. Memo, Gen Arnold to Gen Spaatz, 21 Dec 45, subj: Establishment of Permanent Scientific Advisory Group.

6. Memo, M/Gen LeMay to L/Gen I. C. Eaker, Dep Comdr, AAF, 13 Feb 46, subj: Transfer of Responsibilities and Job Vacancies of the Scientific Advisory Group to the Office DC/AS R&D.

7. Memo, M/Gen LeMay to Gen Spaatz, 3 Jan 46, subj: Establishment of a Permanent Scientific Advisory Group.

8. Memo, Dr. von Karman to CG AAF, 9 Jan 46, subj: Organization of the AAF Scientific Advisory Group.

9. Memo, Gen LeMay to L/Gen Eaker, 13 Feb 46, subj: Transfer of Responsibilities and Job Vacancies of the Scientific Advisory Group to the Office DC/AS R&D.

10. Memo, L/Gen Eaker to M/Gen LeMay, 19 Feb 46, subj: Transfer of Responsibilities and Job Vacancies of the Scientific Advisory Group to the Office DC/AS R&D; Hq AAF HOI 20-76, 4 Mar 46, subj: Organization-The AAF Scientific Advisory Group.

11. Memo, L/Gen Eaker to M/Gen LeMay, 14 Mar 46.

12. Min of SAB Mtg, 18 Jun 46.

13. *Ibid.*

14. *Ibid.*

15. *Ibid.*

16. *Ibid.*

17. *Ibid.*; Memo, M/Gen LeMay to CG AAF, subj: The Scientific Advisory Board; Ltr, M/Gen LeMay to Dr. P. Duwez, 9 May 46.

18. Ltr, Dr. von Karman to M/Gen LeMay, 19 Aug 46 subj: Transmittal of Report on Findings and Recommendations of the SAB.

19. Memo, M/Gen LeMay to CG AAF, 29 Aug 46, subj: Report of SAB, June 1946: Memo, Gen Spaatz to M/Gen LeMay, 4 Sep 46, subj: Report of SAB, June 1946; Memo, M/Gen LeMay to AC/AS-4, 19 Sep 46, subj: Implementation of Recommendations of SAB.

Notes To Pages 19-27.

20. Memo, Dr. von Karman to M/Gen LeMay, 24 Dec 46, subj: SAB Activities and Staff for SAB Office.

21. Memo, Dr. R. P. Johnson to M/Gen LeMay, 27 Dec 46, subj: Plan for Operation of SAB; Hq AAF R&R Sheet, M/Gen LeMay to AC/AS-4, 3 Apr 46, subj: Plan for Operation of SAB.

22. *Ibid.*, Dr. Johnson to M/Gen LeMay, 27 Dec 46.

23. Hq AAF R&R Sheet, M/Gen LeMay to AC/AS-1, 18 Feb 47, subj: Assignment of Officer as Secretary to the SAB.

24. Memo, Dr. von Karman to M/Gen LeMay, 20 May 47, subj: Report of the SAB to the CG AAF.

25. Memo, M/Gen LeMay to Gen Spaatz, 10 Jun 47, subj: Report of the SAB Meeting of February 1947.

26. Ltr, Dr. von Karman to Maj D. M. Alexander, SAB Secy, 15 Sep 47, subj: SAB Meeting.

27. Memo, Dr. Johnson to Col R. C. Wilson, 7 Feb 47.

28. Ltr, Dr. von Karman to Maj Alexander, 15 Sep 47.

29. Hq USAF HOI 20-76, 10 Oct 47, subj: Organization-The USAF Scientific Advisory Board.

30. Ltr, M/Gen L. C. Craigie, Hq USAF Dir R&D, to CG AMC, 7 Jan 48, subj: Problems for Consideration of the SAB; Ltr, M/Gen F. O. Carroll, Hq AMC Dir R&D, to M/Gen Craigie, 11 Mar 48, subj: Problems for Consideration of the SAB.

31. Ltr, M/Gen Craigie to Dr. von Karman, 15 Jan 48; Ltr, Dr. von Karman to M/Gen Craigie, 19 Feb 48.

32. Excerpts, Min of SAB Mtg, 17-18 Mar 48.

33. SAB Memo for Rcrd, 24 May 48, subj: SAB Committees; Memo, Dr. von Karman to CSAF, 7 Jul 48, subj: Report of SAB Meeting, 17-18 Mar 48; Memo, M/Gen Craigie to CSAF, 8 Jul 48, subj: Report of SAB Meeting, 17-18 Mar 48.

34. Memo, Maj Alexander to Dir R&D, 24 Mar 48.

35. Excerpts, Min of SAB Mtg, 17-18 Mar 48.

36. Ltr, M/Gen Craigie to CG AMC, 27 Apr 48, subj: SAB Meetings.

37. Memo, M/Gen Craigie to CSAF, 30 Mar 48, subj: Location of SAB in Air Force Headquarters Organization; Memo, Maj Alexander to Dir R&D 9 Feb 48, subj: Operation and Utilization of the SAB.

38. Ltr, Dr. von Karman to Gen Spaatz, 6 Apr 48.

39. Memo for Rcrd, Mrs. M. D. Roddenberry, SAB Admin Asst, 15 Apr 48, subj: SAB.

40. Memo, M/Gen W. F. McKee, Asst VCS, to Air Stf, 26 Apr 48, subj: The SAB to the Chief of Staff, USAF.

CHAPTER THREE

1. Hq AAF R&R Sheet, M/Gen Craigie to Secy of Air Stf, 26 May 48, subj: Proposed Transfer of the SAB Secretariat; Memo, SAB Admin Asst to M/Gen Craigie, 28 May 48, Manning Table for the SAB Secretari-

Notes To Pages 28-36.

at; Memo, M/Gen Craigie to Secy of Air Stf, 23 Jul 48, subj: Civilian Personnel Authorization-SAB Secretariat.

2. Memo, Maj T. F. Walkowicz, SAB Secy, to SAB Chmn and Mil Dir, 10 Nov 48, subj: Effective Utilization of the SAB.

3. Min of SAB Mtg, 18 Nov 48.

4. Memo, Maj Walkowicz to SAB Chmn and Mil Dir, 10 Nov 48.

5. *Ibid.*

6. Memo, Gen H. S. Vandenberg, CSAF, to Air Stf, 26 May 49, subj: Implementation of SAB Recommendations.

7. *Ibid.*

8. Ltr, B/Gen D. L. Putt, Dir R&D, to CG AMC, 22 Oct 48, subj: AMC SAB Office.

9. Min of SAB Mtg, 18 Nov 48.

10. Ltr, L/Gen B. W. Chidlaw, CG AMC, to Dr. von Karman, 17 Dec 48.

11. *Ibid.*

12. Ltr, Dr. von Karman to L/Gen Chidlaw, 13 Jan 49.

13. Min of SAB Mtg, 17 Nov 48.

14. Ltr, Dr. von Karman to Gen Vandenberg, 15 Jan 49.

15. M/Gen D. L. Putt, SAB Mil Dir, Min of Remarks to Stf Mtg of Engineering Div, AMC, 4 Apr 49.

16. Min of SAB Mtg, 7 Apr 49.

17. *Ibid.*

18. *Ibid.*

19. Ltr, Dr. von Karman to Gen Vandenberg, 14 Apr 49.

20. Excerpts, Min of SAB Study Gp Mtg, 11 Jul 49.

21. *Ibid.*

22. Min of SAB Mtg, 12 Apr 50.

23. Min of SAB Mtg, 3 Nov 49.

24. *Ibid.*

25. *Ibid.*

26. *Ibid.*

27. Ltr, Dr. von Karman to Gen Vandenberg, 21 Sep 49.

28. Min of SAB Mtg, 3 Nov 49.

29. Memo, Gen M. S. Fairchild, VCS, to Air Stf, 23 Jan 50, subj: Organization for Research & Development in the USAF.

30. Ltr, M/Gen McKee (for Gen Vandenberg) to SAB Chmn, 24 May 50, subj: Survey of Air Force Medical Research.

31. Min of SAB Ad Hoc Cmte on the Review of Medical Research Meeting, 9 Oct 50.

32. Memo, SAB Secy to AF Hist Office, 30 Jun 51, subj: Review of SAB FY 1951 Activities (hereafter referred to as SAB Hist Rprt, by appropriate year).

33. SAB Hist Rprt, FY 1950.

Notes To Pages 39-46.

CHAPTER FOUR

1. Min of SAB Exec Cmte Mtg, 2 Nov 49.

2. Progress Rprt, ADSEC, 1 May 50; SAB Hist Rprt, FY 1950.

3. Memo, Dr. von Karman to Gen Fairchild, 29 Nov 49.

4. SAB Hist Rprt, FY 1950.

5. SAB Hist Rprt, FY 1952.

6. ADSEC Progress Rprt, 1 May 50.

7. Remarks to SAB Mtg by Gen N. F. Twining, CSAF, 19 Oct 53.

8. SAB Hist Rprt, FY 1951.

9. Ltr, Gen Vandenberg to Dr. von Karman, 11 Feb 42; Memo, M/Gen S. E. Anderson, Dir Plans & Ops, to M/Gen G. P. Saville, DCS/D, 17 Jul 50.

10. Rprt of Special SAB Cmte (Draft), 29 Jul 50, subj: Recommendations on Air Force Guided Missile Program; SAB Hist Rprt, FY 1951.

11. Memo, RDB 265/26, 25 Aug 50, subj: Approval for Additional Support for the FY 1951 R&D Program.

12. Memo, M/Gen Saville to Dir R&D, 8 Sep 50, subj: Air Force Supplemental FY 1951 R&D Program.

13. Reports of Panels and Systems Cmtes, SAB Mtg, 11-15 Sept 50.

14. Memo, Dr. von Karman to Gen Vandenberg, 18 Sep 50.

15. SAB Hist Rprt, FY 1951.

16. Min of SAB Mtg, 13 Sep 50.

17. *Ibid.*

18. Memo, Gen Vandenberg to VCS, 12 Oct 50, subj: Organization for Research and Development in the USAF.

19. Memo, Dr. L. N. Ridenour to Dr. von Karman, 7 May 51, subj: Interim Report on Activities of the Special Working Group, SAB; Memo for Rcrd, Dr. Ridenour, 7 Aug 51, subj: Minutes of the 6th July 1951 Meeting of the Special Working Group of the SAB.

20. Ltr, Dr. J. H. Doolittle to Gen Twining, 10 Mar 54.

21. Memo, Gen Vandenberg to Dr. von Karman, 8 May 51, subj: Request for Study of Air Force Armament Activities; Memo for Record, Mr. B. J. Driscoll, SAB Secy, 14 May 51, subj: Executive Committee Meeting, 2 May 51.

22. SAB Hist Rprt, FY 1951.

23. Min of SAB Ad Hoc Cmte on Armament and Ordnance Mtg, 15-19 Jun 51.

24. Rprt of SAB Ad Hoc Cmte, 4 Oct 51, subj: Armament and Ordnance Research and Development in the USAF.

25. SAB Hist Rprt, FY 1952.

26. *Ibid.*

27. Memo, M/Gen McKee (for Gen Vandenberg) to Air Stf, 4 Oct 50, subj: Implementation of SAB Recommendations; Memo, Gen Twining (for Gen Vandenberg) to Air Stf, 22 May 51, subj: Implementation of SAB Recommendations.

Notes To Pages 69-81.

8. Min of SAB Exec Cmte Mtg, 14 Nov 56.

9. Memo, Prof. C. D. Perkins, AF Chief Scientist, to Gen Twining, 8 Apr 57, subj: Rpt of Committee to Study the Activities of the SAB.

10. *Ibid.*

11. Min of SAB Exec Cmte Mtg, 10 Apr 56.

12. Memo, Mr. Hasert to SAB Panel Chairmen, 12 Mar 57; Min of SAB Exec Cmte Mtg, 19 May 57.

13. Ltr, Dr. C. S. White to Dr. Doolittle, 13 May 57.

14. Ltr, Dr. C. B. Millikan to SAB Chairman, 4 Apr 57, subj: SAB Aircraft Panel Chairman's Report to the May 1957 SAB Meeting.

15. Dr. I. A. Getting, May 57, subj: Electronics and Communications Panel Activities in Recent Years.

16. *Ibid.*

17. *Ibid.*

18. Dr. C. S. Draper, 30 Apr 57, subj: Review of Explosives and Armament Panel Activities.

19. *Ibid.*

20. *Ibid.*

21. Dr. M. M. Mills, 2 May 57, subj: Review of Activities of the Fuels and Propulsion Panel.

22. Ltr, Dr. J. Kaplan to Dr. Doolittle, 19 Apr 57.

23. Ltr, L/Gen T. D. White, Actg VCS, to Dr. von Karman, 8 Jun 53.

24. Min of SAB Exec Cmte Mtg, 21 Oct 53.

25. Ltr, Mr. Hasert to Dr. von Karman, 19 Aug 53; Memo, Gen Putt to Col R. J. Burger, SAB Secretary, 18 Nov 65.

26. Memo, Chairman of Nuclear Panel, May 57, subj: Activities of the Nuclear Panel.

27. Min of SAB Exec Cmte Mtg, 21 Oct 53.

28. *Ibid.*

29. Memo, Dr. D. E. Macdonald to SAB Chmn, 1 May 57, subj: Activities of the Reconnaissance Panel.

30. Memo, Dr. D. Wolfle to Dr. Doolittle, 2 May 57, subj: Work of the SAB Social Sciences Panel.

CHAPTER EIGHT

1. Dr. L. Bowen, USAF Historical Division Study, *An Air Force History of Space Activities*, 1945-1959, pp 10-12.

2. Ltr, L/Col B. C. Gray, SAB Asst Secy, to Dr. Doolittle, 26 Nov 57.

3. Memo, Mr. Hasert to SAB Mil Dir, 11 Dec 57; Memo for Rcrd, Mr. Hasert, 17 Oct 57, subj: SAB Space Technology Activities.

4. Hq ARDC, 28 Oct 57, subj: Report of the Teller Ad Hoc Cmte.

5. Memo, Col T. Drysdale, Chief Research & Analysis Div, DCS/D, to Asst for Development Programs, 26 Nov 57, subj: Report of Teller Ad Hoc Cmte.

6. Rprt, SAB Fuels and Propulsion Panel, 14 Nov 56.

Notes To Pages 57-68.

4. Min of SAB Exec Cmte Mtg, 16 Jun 54.

5. Ltr, Mr. Hasert to Maj Whitcraft, 3 Dec 53; Ltr, L/Gen Craigie to L/Gen Putt, 9 Dec 53.

6. Memo, Dr. von Karman to SAB Panel Chairmen, 29 Jan 54, sub: *Toward New Horizons* Revision Requested by Gen. Putt.

7. Dr. I. A. Getting, 23 Mar 54, subj: Preliminary Report of the Electronics and Communications Panel.

8. Dr. D. W. Hastings, 1 Oct 54, subj: Some Remarks of the Aeromedical Panel on New Developments of the Next Ten Years; SAB Hist Rpt, FY 1954.

9. Dr. C. B. Millikan, 1 Oct 54, subj: Some remarks of the Aircraft Panel on New Technical Developments of the Next Ten Years.

10. Dr. Getting, 23 Mar 54.

11. Dr. R. H. Kent, 23 Mar 54, subj: Preliminary Report of the Explosives and Armament Panel on New Horizons.

12. Mr. A. M. Rothrock, 23 Mar 54 subj: Preliminary Report of the Fuels and Propulsion Panel on New Horizons.

13. Dr. H. Wexler, 23 Mar 54, subj: Preliminary Report of the Geophysical Panel on New Horizons.

14. Dr. J. von Neumann, 23 Mar 54, subj: Preliminary Report of the Nuclear Weapons Panel on New Horizons.

15. Dr. J. G. Baker, 23 Mar 54, subj: Preliminary Report of the Intelligence Systems Panel on New Horizons.

16. Dr. J. W. Gardner, 23 Mar 54, subj: Preliminary Report of the Social Sciences Panel on New Horizons.

17. SAB Hist Rprt, FY 1954.

18. Min of SAB Exec Cmte Mtg, 23 Mar 54.

19. Min of SAB Exec Cmte Mtg, 16 Jun 54.

20. *Ibid.*

21. Min of SAB Exec Cmte Mtg, 26 Sep 54.

22. Ltr, Dr. Doolittle to L/Gen Putt, 29 Sep 54.

23. *Ibid.*

CHAPTER SEVEN

1. Ltr, Dr. von Karman to Gen Twining, 17 Sep 54; Memo, Gen Twining to L/Gen Putt, 20 Dec 54; Ltr, Gen Twining to Dr. von Karman, 31 Dec 54.

2. Min of SAB Policy Gp Mtg, 21 Mar 54.

3. Min of SAB Exec Cmte Mtg, 23 Mar 55.

4. Ltr, Dr. Doolittle to Dr. M. J. Kelly, 21 Feb 57.

5. Min of SAB Exec Cmte Mtg, 12 Jun 55; Memo, L/Gen Putt to CSAF, 21 Oct 54, subj: Relative Rank of SAB Members; Ltr, Gen Twining to SAB Mil Dir, 27 Oct 54, subj: Relative Rank of SAB Members.

6. SAB Hist Rprt (Draft), FY 1955.

7. Min of SAB Exec Cmte Mtg, 10 Apr 56.

8. Min of SAB Exec Cmte Mtg, 14 Nov 56.

9. Memo, Prof. C. D. Perkins, AF Chief Scientist, to Gen Twining, 3 Apr 57, subj: Rpt of Committee to Study the Activities of the SAB.

10. *Ibid.*

11. Min of SAB Exec Cmte Mtg, 10 Apr 56.

12. Memo, Mr. Hasert to SAB Panel Chairmen, 12 Mar 57; Min of SAB Exec Cmte Mtg, 19 May 57.

13. Ltr, Dr. C. S. White to Dr. Doolittle, 13 May 57.

14. Ltr, Dr. C. B. Millikan to SAB Chairman, 4 Apr 57, subj: SAB Aircraft Panel Chairman's Report to the May 1957 SAB Meeting.

15. Dr. I. A. Getting, May 57, subj: Electronics and Communications Panel Activities in Recent Years.

16. *Ibid.*

17. *Ibid.*

18. Dr. C. S. Draper, 30 Apr 57, subj: Review of Explosives and Armament Panel Activities.

19. *Ibid.*

20. *Ibid.*

21. Dr. M. M. Mills, 2 May 57, subj: Review of Activities of the Fuels and Propulsion Panel.

22. Ltr, Dr. J. Kaplan to Dr. Doolittle, 19 Apr 57.

23. Ltr, L/Gen T. D. White, Actg VCS, to Dr. von Karman, 8 Jun 53.

24. Min of SAB Exec Cmte Mtg, 21 Oct 53.

25. Ltr, Mr. Hasert to Dr. von Karman, 19 Aug 53; Memo, Gen Putt to Col R. J. Burger, SAB Secretary, 18 Nov 65.

26. Memo, Chairman of Nuclear Panel, May 57, subj: Activities of the Nuclear Panel.

27. Min of SAB Exec Cmte Mtg, 21 Oct 53.

28. *Ibid.*

29. Memo, Dr. D. E. Macdonald to SAB Chmn, 1 May 57, subj: Activities of the Reconnaissance Panel.

30. Memo, Dr. D. Wolfle to Dr. Doolittle, 2 May 57, subj: Work of the SAB Social Sciences Panel.

CHAPTER EIGHT

1. Dr. L. Bowen, USAF Historical Division Study, *An Air Force History of Space Activities,* 1945-1959, pp 10-12.

2. Ltr, L/Col B. C. Gray, SAB Asst Secy, to Dr. Doolittle, 26 Nov 57.

3. Memo, Mr. Hasert to SAB Mil Dir, 11 Dec 57; Memo for Rcrd, Mr. Hasert, 17 Oct 57, subj: SAB Space Technology Activities.

4. Hq ARDC, 28 Oct 57, subj: Report of the Teller Ad Hoc Cmte.

5. Memo, Col T. Drysdale, Chief Research & Analysis Div, DCS/D, to Asst for Development Programs, 26 Nov 57, subj: Report of Teller Ad Hoc Cmte.

6. Rprt, SAB Fuels and Propulsion Panel, 14 Nov 56.

Notes To Pages 82-90.

7. Memo, L/Gen Putt to SAB Chmn, 15 May 57, subj: SAB Special Study of Advanced Weapons Technology and Environment.

8. Rprt of the SAB Ad Hoc Cmte on Advanced Weapons Technology and Environment, 9 Oct 57; Memo, Mr. Hasert to SAB Mil Dir, 11 Dec 57.

9. Memo, Dr. Doolittle to CSAF, 9 Oct 57, subj: Report of the SAB Ad Hoc Cmte on Advanced Weapons Technology and Environment.

10. Memo, Mr. Hasert to SAB Mil Dir, 11 Dec 57.

11. Memo, Dr. Doolittle to Gen T. D. White, CSAF, 9 Dec 57, subj: Space Technology.

12. Ltr, L/Gen Putt to SAB Chmn, 20 Dec 57, subj: SAB Study Group.

13. Ltr, Col G. H. Duncan, SAB Secy, to Prof. C. W. Sherwin, 14 Jan 58; Ltr, Col Duncan to Dr. Doolittle, 15 Aug 58.

14. Memo, L/Gen R. C. Wilson, SAB Mil Dir, to SAB Chmn, 22 Aug 58, subj: Amendment to Memo to SAB Chairman (20 Dec 57, subj: SAB Study Group).

15. Final Rprt of SAB Ad Hoc Cmte on Air Defense Systems, 15 Jan 59.

16. Ltr, L/Gen Putt to Prof. Sherwin, 4 Jun 59.

17. Memo, Gen White to SAB Chmn, 21 Nov 57, subj: Review of Air Force R&D Accomplishments; Mr. S. Milner, Hq OAR Hist Div, *AFRD to OAR: An Organizational and Administrative History* (Draft), pp 9-19.

18. *Ibid.*, Milner.

19. Rprt of the SAB Ad Hoc Cmte on Research and Development, June 1958.

20. *Ibid.*

21. *Ibid.*

22. Ltr, L/Gen Wilson to Gen S. E. Anderson, ARDC Comdr, 24 Jun 58, Report of the SAB Ad Hoc Cmte on Research and Development.

23. Hq ARDC Staff Study, 31 Jul 58, subj: Report of the SAB Ad Hoc Cmte on Research and Development; Hq USAF Staff Study, Aug 58, subj: Evaluation of the June 1958 SAB Report on R&D.

24. Milner, *AFRD to OAR: An Organizational and Administrative History* (Draft), p 76.

25. Memo, Dr. Doolittle to Panel Chairmen, 26 Nov 57, subj: Future Panel Activities.

26. Ltr, Dr. E. H. Plesset to Col Duncan, 30 Dec 57.

27. Min of SAB Exec Cmte Mtg, 3 Dec 57.

CHAPTER NINE

1. Ltr, Dr. Doolittle to Gen White, 7 Oct 58; Ltr, White to Doolittle, 27 Oct. 58.

2. Ltr, Prof. C. D. Perkins to L/Gen Putt, 16 Mar 59, subj: Reorganization of SAB.

3. Ltr, Prof. Perkins to Col Duncan, 12 Sep 58, with incl: 12 Sep 58, subj: Preliminary Report of the Ad Hoc Panel on Realignment of the SAB Membership; Min of SAB Exec Cmte Mtg, 29 Apr 58.

4. *Ibid.,* Prelim Rprt, 12 Sep 58.

5. Memo, Col Duncan to Asst DCS/D, 30 Sep 58, subj: Fall Meeting, SAB.

6. Memo, Mr. Hasert to SAB Exec Cmte Members, 31 Oct 58, subj: SAB Organization; Ltr, Dr. Stever to Col Gasser, 5 Mar 59.

7. Memo for Rcrd, Col C. D. Gasser, SAB Secy, 18 Dec 58.

8. Ltr, Col Gasser to Dr. H. G. Stever, 6 Feb 59; Ltr, Col Gasser to L/Gen Putt, SAB Chmn, 27 Feb 61.

9. Ltr, Dr. Macdonald to Col Gasser, 26 Mar 59.

10. Ltr, Prof. Perkins to L/Gen Putt, 16 Mar 59.

11. Ltr, Dr. Stever to Col Gasser, 5 Mar 59.

12. Ltr, Dr. C. B. Millikan to Dr. Doolittle, 31 Dec 57; Ltr, Dr. Doolittle to Dr. Millikan, 16 Jan 58, Memo, Dr. Doolittle to SAB Members, 19 Dec 57, subj: Senior Statesmen.

13. Ltr, L/Gen Putt to CSAF, [ca.] Dec 58.

14. Memo, Mr. Hasert to SAB Mil Dir, 11 Dec 57.

15. Ltr, Dr. Doolittle to Gen White, 10 Jul 58.

16. Memo for Rcrd, Col Gasser, 23 Dec 58.

17. Ltr, Mr. Hasert to L/Col D. L. Carter, Hq ARDC, 13 Feb 59.

18. Ltr, L/Gen Putt to L/Gen Wilson, 9 Apr 59, subj: Space Technology Problem Areas.

19. Memo, Col Gasser to Dr. Stever, 29 May 59, subj: DCS/D Position on Space Technology Problem Areas.

20. Memo, L/Gen Wilson to SAB Chmn, 25 May 59, subj: Space Technology Problem Areas.

21. Memo for Rcrd, Col Gasser, 29 Jul 59.

22. Memo, Mr. Hasert to Space Technology Panel Members, 7 Jan 60, subj: Meeting 28-30 Jan 60.

23. Summary Rprt of SAB Space Technology Panel, 30 Dec 60, subj: Space Technology—December 1960.

24. Ltr, L/Gen Putt to Dr. G. E. Valley, 2 Jul 59; Memo, Col Gasser to SAB Exec Cmte Members, 13 Jul 59; Memo, Col Gasser to SAB Exec Cmte and Basic Research Panel Members, 13 Oct 59, subj: Panel Charter, with incl: Dr. Valley, 9 Oct 59, subj: Terms of Reference for the SAB Panel on Basic Research; Min of First Meeting of Basic Research Panel, 7 Dec 59.

25. Ltr, B/Gen B. G. Holzman, AFRD Comdr, to L/Gen Putt, 14 Jan 60.

26. Ltr, L/Gen Putt to Dr. Stever, 21 Feb 61, with incl: Dr. Valley, 13 Feb 61, subj: Report of the SAB Basic Research Panel (Draft), February 7-8, 1961.

27. Ltr, Gen C. E. LeMay, CSAF, to L/Gen Putt, 3 Nov 61; Mr. C. Berger, USAF Historical Div Study, 3 Dec 62, *The Strengthening of Air Force In-House Laboratories, 1961-1962.*

Notes To Pages 98-105.

28. Rprt of SAB Ad Hoc Cmte on Technical Facilities, Feb 62.

29. Rprt of SAB Ad Hoc Cmte on Effectiveness of USAF In-House Laboratories Organization and Manning, Apr 62.

30. M/Gen M. C. Demler, RTD Comdr, in address at SAB Mtg, 11 Apr 63.

31. Ltr, L/Gen Wilson to SAB Chmn, 19 Feb 60, subj: Review of Research Programs on Large Capacity Computers.

32. Rprt of SAB Electronics Panel, 26 Apr 60, subj: Air Force Communications Problems; Ltr, L/Gen Wilson to SAB Chmn, Dec 60, subj: Air Force Problem Areas for SAB Consideration in Calendar Year 1961.

33. *Ibid.*, L/Gen Wilson to SAB Chmn, Dec 60.

34. Rprt of the SAB Ad Hoc Panel on Information Processing, 19 Apr 61; Min of SAB Exec Cmte Mtg, 27 Apr 61; Ltr, Dr. Stever to Info Proces Panel, 1 Jun 62.

35. Ltr, L/Gen Putt to CSAF, 12 Jan 60, subj: SAB Report on Atomic Test Moratorium.

36. Ltr, Gen White to Dr. Doolittle, 20 Jan 60; Ltr, Col Gasser to Dr. C. C. Lauritsen, 8 Feb 60.

37. Ltr, L/Gen Wilson to Gen White, 29 Mar 60; Ltr, Dr. L. F. Carter to Dr. Stever, 5 Feb 62.

38. Memo, Prof. T. C. Schelling to Arms Control Committee, 17 Mar 61, subj: Arms Control and Command Control.

CHAPTER TEN

1. Ltr, L/Gen Putt to Col Gasser, 1 Jul 59.

2. Ltr, L/Gen Putt to SAB Exec Cmte Members, 26 Oct 59.

3. Memo for Rcrd, Col Gasser, 19 Mar 62, subj: SAB Division/Center Advisory Groups to AFSC.

4. Memo, L/Gen Wilson to VCS, 21 May 60.

5. Ltr, Gen B. A. Schriever, ARDC Comdr, to L/Gen Putt, 15 Sep 59.

6. Ltr, Gen Schriever to L/Gen Putt, 21 Oct 60.

7. Memo for Rcrd, Col Gasser, 19 Mar 62.

8. Min of AFSC/SAB Mtg on DAG's, 2 Apr 62.

9. Ltr, Dr. Stever to Gen White, 24 Apr 61.

10. Min of SAB Exec Cmte Mtg, 27 Apr 61; SAB *Diary*, 2 May 61.

11. Min of SAB Exec Cmte Mtg, 2 May 61.

12. Memo for Rcrd, Col Gasser, 19 Mar 62.

13. Min of SAB Exec Cmte Mtg, 19 Jun 61, as recorded in SAB *Diary*, 19 Jun 61.

14. Memo for Rcrd, Col Gasser, 19 Mar 62.

15. Ltr, Col H. F. Bunze, Hq BSD, to SAB, 12 Jun 62, subj: Data on BSD Advisory Group Meeting.

16. Min of AFSC/SAB Meeting on DAG's, 2 Apr 62.

17. Ltr, Dr. A. H. Flax to M/Gen W. A. Davis, ASD Comdr, 18 May 62.

Notes To Pages 105-113.

18. Ltr, Gen Schriever to L/Gen Putt, 4 Aug 61.

19. Rprt of RTAG (Draft), [ca] Aug 61.

20. Ltr, Gen Schriever to L/Gen Putt, 25 Sep 61; Ltr, L/Gen Putt to Gen Schriever, 5 Oct 61.

21. M/Gen M. F. Cooper, Hq AFSC, in Min of RTAG Mtg, 10 Jul 62.

22. Ltr, Mr. A. F. Arcier, FTD (AFSC) Scientific Advisor to Col Gasser, [ca] Jul 62.

23. ESD Reg 20-2, 14 Sep 62, subj: Organization and Mission-General: Division Advisory Group.

CHAPTER ELEVEN

1. Min of SAB Mtg, 16 Jun 46.

2. Memo, Dr. Root to Mr. Driscoll, 11 Aug 52, subj: Suggested Item for Agenda-Executive Committee Meeting of the SAB.

3. SAB Exec Cmte Mtg, Addendum to Agenda, 29 Mar 53.

4. Min of SAB Policy Gp Mtg, 21 Mar 54.

5. Ltr, L/Gen Anderson to L/Gen Putt, 7 Apr 58; Ltr, L/Gen Putt to L/Gen Anderson, 14 Apr 58.

6. Memo for Rcrd, Col Gasser, 15 Dec 58.

7. Ltr, Mr. R. Gilpatric, Dep SOD, to DOD Agencies, 23 Mar 61.

8. Ltr, Gen LeMay, CSAF, to DDR&E, 20 Jul 61.

9. Mr. J. W. Finney, *New York Times*, 30 Dec 61, subj: 2 Defense Roles Laid to General.

10. Memo SAF to SOD, [ca] 20 Jan 62, subj: Recent Appointment Instructions Affecting the USAF Scientific Advisory Board.

11. Ltr, L/Gen J. Ferguson, SAB Mil Dir, to CSAF, 10 Jan 62, subj: Conflict of Interest Matters of Concern to the SAB.

12. *Ibid.*

13. Attorney General To President of the United States, 31 Jan 62.

14. Presidential Executive Order, 26 Feb 62, subj: Prescribing Regulations for the Formation and Use of Advisory Committees.

15. Memo, Col Gasser to SAB Members, 25 Apr 62, subj: Standards of Conduct for Experts and Consultants.

16. Memo, Dr. J. V. Charyk, Under SAF, to Dep SOD, subj: Request for Approval of USAF SAB Personnel Action; Agenda of SAB Exec Cmte Mtg (Tab F), 19 Jun 62.

17. Memo, Col Gasser to SAB Secyt, 16 Mar 62, subj: Procedure to be Followed Relative SAB Operations.

18. Memo, Mr. R. S. McNamara, SOD, to Mr. E. M. Zuckert, SAF, 21 Jun 62; Memo, Mr. Zuckert to Gen LeMay, 22 Jun 62; Memo, L/Gen Ferguson to Secy Air Stf, 4 May 64; subj: Advisory Committee; Memo, Gen LeMay to Mr. Zuckert, 3 Jul 62, subj: SOD Directive Relative to the USAF Scientific Advisory Board dated 21 June 1962; Memo, Mr. Zuckert to Mr. McNamara, 17 Jul 62, subj: Air Force Scientific Advisory Board.

19. Memo, L/Gen Ferguson to Secy Air Stf, 4 May 64.

Notes To Pages 113-121.

20. Msg 72524, SAB Secy to SAB Panel Chairmen, 6 Aug 62.

21. Ltr, Dr. Stever to L/Gen Ferguson, 29 Jan 62.

22. Ltr, L/Gen Ferguson to L/Gen H. M. Estes, V/Comdr ARDC, 15 Mar 63.

23. Ltr, SAB Secy to DAG Chairmen, 19 Sep 62, in Agenda to Exec Cmte Mtg (TAB E), 27 Oct 62.

24. Min of ESD DAG Mtg of 1-2 Nov 62, 28 Nov 62.

25. Ltr, Gen Schriever to B/Gen A. J. Pierce, FTD Comdr, 28 Nov 62.

26. Ltr, Dr. Stever to Gen Schriever, 26 Feb 63; Memo, Col Gasser to SAB Members, 17 Apr 63, subj: Use of Members of Non-Profit Organizations on SAB-Sponsored Division Advisory Groups; Ltr, Gen Schriever to AFSC Div Comdrs, 3 May 63: Ltr, L/Gen Ferguson to L/Gen Estes, 15 Mar 63.

27. Min of SAB Exec Cmte Mtg, 12 Apr 63.

28. Ltr, Gen Schriever to AFSC Div Comdrs, 3 May 63.

29. Min of SAB Exec Cmte Mtg, 21 Jan 64.

30. Ltr, Mr. Zuckert to Dr. Stever, 27 Aug 64, appended to SAB Annual Report, 1964, p 40.

CHAPTER TWELVE

1. SAB *Annual Reports*, 1959-1964

2. Ltr, Dr. Flax, AF Chief Scientist, to Gen LeMay, 15 Mar 61, subj: Summary of Discussion on Effectiveness of Scientific Support of the Air Force by the SAB.

3. Ltr, Gen White to SAB Chmn, 28 Feb 61; Memo, Col Gasser to L/Gen Putt, 28 Feb 61, subj: Request for SAB Views on Ways and Means to Best Use Scientific Talent Within and Outside the USAF.

4. Memo for Rcrd, Mr. Hasert, 15 Mar 61, subj: Dr. Stever's Conference with Gen White, 15 Mar 61.

5. Rprt of the SAB Cmte on Air Force Utilization of Scientific Resources, 26 May 61.

6. *Ibid.*

7. Memo, Col Gasser to Panel Chairmen, 20 Feb 61, subj: Improved USAF SAB Operational Procedure.

8. SAB *Diary*, 11 Jan 62.

9. Ltr, Dr. Stever to Gen LeMay, 11 Apr 62.

10. SAB *Diary*, 11 Jan 62.

11. Min of SAB Exec Cmte Mtg, 23 Jan 63; Ltr, L/Gen Ferguson to M/Gen D. R. Ostrander, OAR Comdr, 26 Jun 63.

12. Min of SAB Exec Cmte Mtg, 21 Jan 64.

13. Ltr, Dr. Flax to Gen LeMay, 15 Mar 61.

14. Min of SAB Exec Cmte Mtg (TAB C, subj: Future Role of Board, 23 Jan 63.

15. Mr. J. H. Rubel, Dep DDR&E, in address at SAB Mtg, 25 Oct 62.

Notes To Pages 121-130.

16. SAB Memo for Rcrd, 17 Dec 62, subj: SAB Long-Term Objectives and Structure.

17. Dr. Flax, Report on Results of Organization Investigation (TAB C to Min of SAB Exec Cmte Mtg, 24 Apr 62.)

18. R/Adm (Ret) P. A. Smith, 23 Oct 62, subj: Some Considerations Regarding SAB Structure and Working Methods.

19. Memo, L/Gen Ferguson to DCS/R&T Stf, 19 Jul 63.

20. Min of SAB Steering Cmte, 12 Dec 63.

21. Min of SAB Exec Cmte Mtg, 21 Jan 64.

22. *Ibid.*

23. *Ibid.*

24. *Ibid.*

25. Min of SAB Steering Cmte Mtg, 28 Feb 64.

26. Ltr, Dr. L. S. Sheingold to Dr. H. M. Agnew, 2 Mar 64.

27. Ltr, Dr. Stever to Members SAB Exec Cmte, 30 Mar 64.

28. Memo, Col R. J. Burger, SAB Secy, to SAB Panel Members, 13 Apr 64, subj: SAB Task Groups.

29. Ltr, Gen LeMay to Dr. Stever, 15 May 64.

30. Min of SAB Exec Cmte Mtg, 31 May 64.

31. Ltr, Dr. Sheingold to Gen W. C. Sweeney, TAC Comdr, 23 Jun 64.

32. Msg 89891, SAB Secyt to Working Gp Chairmen, 23 Jun 64.

33. Ltr, Gen Schriever to Dr. Sheingold, 3 Aug 64.

34. Notes of Mtg, Mr. Zuckert with L/Gen Ferguson and Dr. Sheingold, 25 Aug 64.

35. Ltr, M/Gen Ostrander to Col Burger, 23 Sep 64.

36. Memo, Dr. Sheingold to SAB Panel and TAC Task Force Working Gp Chairmen, 23 Sep 64.

37. Notes of Mtg, L/Gen Ferguson and Dr. Sheingold with Mr. Zuckert, 25 Aug 64.

38. *Ibid.*

39. Min of SAB Exec Cmte Mtg, 30 Oct 64.

40. Ltr, Mr. Zuckert to Dr. Stever, 29 Jul 64.

GLOSSARY

AAF	Army Air Forces
AC/AS	Assistant Chief of Air Staff
Actg	Acting
ADSEC	Air Defense Systems Engineering Committee
Advsy	Advisory
AEC	Atomic Energy Commission
AEDC	Arnold Engineering Development Center
AF	Air Force
AFCRL	Air Force Cambridge Research Laboratories
AFRD	Air Force Research Division
AFSC	Air Force Systems Command
AGARD	Advisory Group for Aeronautical Research and Development
AMC	Air Materiel Command
AMR	Atlantic Missile Range
ARDC	Air Research and Development Command
ASD	Aeronautical Systems Division
Asst	Assistant
BSD	Ballistic Systems Division
CG	Commanding General
Chmn	Chairman
Cmte	Committee
Comdr	Commander
Con	Control
Coun	Council
CSAF	Chief of Staff, U.S. Air Force
DAG	Division Advisory Group
DC/AS	Deputy Chief of Air Staff
DCS/D	Deputy Chief of Staff, Development
DCS/R&T	Deputy Chief of Staff, Research and Technology
DDR&E	Director, Defense Research and Engineering
Dep	Deputy
Dir	Director
Div	Division
DOD	Department of Defense
ESD	Electronic Systems Division
Exec	Executive
FTD	Foreign Technology Division
FY	Fiscal Year
Gp	Group
Hist	History
HOI	Headquarters Office Instruction
Hq	Headquarters

ICBM	Intercontinental Ballistic Missile
Incl	Inclosure
Ltr	Letter
Mgt	Management
Mil	Military
Min	Minutes
Mtg	Meeting
NASA	National Aeronautics and Space Administration
NATO	North Atlantic Treaty Organization
NSF	National Science Foundation
Ops	Operations
OSD	Office of the Secretary of Defense
OSR	Office of Scientific Research
Plcy	Policy
QRC	Quick Reaction Capability
R&D	Research and Development
RTD	Research and Technology Division
R&R	Routing and Record (Sheet)
Rcrd	Record
RDB	Research and Development Board
Reqmts	Requirements
Rprt	Report
RTAG	Range Technical Advisory Group
SAB	Scientific Advisory Board
SAF	Secretary of the Air Force
SAFOI	Secretary of the Air Force, Office of Information
SAG	Scientific Advisory Group
SOD	Secretary of Defense
Secy	Secretary
Secyt	Secretariat
SSD	Space Systems Division
Stf	Staff
Subj	Subject
Sup	Supplement
Tech	Technical
USAF	United States Air Force
VCS	Vice Chief of Staff

www.ingramcontent.com/pod-product-compliance
Lightning Source LLC
Chambersburg PA
CBHW080806180526
45168CB00006B/2342